SURFACE CHEMISTRY
and GEOCHEMISTRY of
HYDRAULIC FRACTURING

SURFACE CHEMISTRY
and GEOCHEMISTRY of
HYDRAULIC FRACTURING

K.S. BIRDI

CRC Press
Taylor & Francis Group
Boca Raton   London   New York

CRC Press is an imprint of the
Taylor & Francis Group, an **informa** business

CRC Press
Taylor & Francis Group
6000 Broken Sound Parkway NW, Suite 300
Boca Raton, FL 33487-2742

First issued in paperback 2020

© 2017 by Taylor & Francis Group, LLC
CRC Press is an imprint of Taylor & Francis Group, an Informa business

No claim to original U.S. Government works

ISBN-13: 978-1-4822-5718-2 (hbk)
ISBN-13: 978-0-367-51600-0 (pbk)

Library of Congress Cataloging-in-Publication Data

Names: Birdi, K. S., 1934-
Title: Surface chemistry and geochemistry of hydraulic fracturing / K.S. Birdi.
Description: Boca Raton : Taylor & Francis Group, 2017. | "A CRC title." | Includes bibliographical references and index.
Identifiers: LCCN 2016014465 | ISBN 9781482257182 (alk. paper)
Subjects: LCSH: Hydraulic fracturing. | Hydraulic fracturing--Environmental aspects. | Geochemistry. | Surface chemistry. | Surface tension. | Gases--Absorption and adsorption.
Classification: LCC TN871.255 .B57 2017 | DDC 622/.3381--dc23
LC record available at https://lccn.loc.gov/2016014465

Visit the Taylor & Francis Web site at
http://www.taylorandfrancis.com

and the CRC Press Web site at
http://www.crcpress.com

Dedication

To Leon, Esma and David

Contents

Author xi

Chapter 1 Surface Chemistry and Geochemistry of Hydraulic Fracturing 1

    1.1   Introduction 1
    1.2   Formation of Fractures in Shale Reservoirs and Surface
           Forces 7
    1.3   Colloids 17
    1.4   Emulsions (and Hydraulic Fracking Fluids) 18

Chapter 2 Capillary Forces in Fluid Flow in Porous Solids (Shale
                  Formations) 21

    2.1   Introduction 21
    2.2   Surface Forces in Liquids 23
           2.2.1   Surface Energy 24
    2.3   Laplace Equation for Liquids (Liquid Surface Curvature
           and Pressure) 27
    2.4   Capillary Rise (or Fall) of Liquids 33
    2.5   Bubble (or Foam) Formation 36
    2.6   Measurement of Surface Tension of Liquids 38
           2.6.1   Liquid Drop Weight and Shape Method 38
                     2.6.1.1   Maximum Weight Method 40
                     2.6.1.2   Shape of the Liquid Drop (Pendant
                                Drop Method) 40
           2.6.2   Plate Method (Wilhelmy) 41
    2.7   Surface Tension Data of Some Typical Liquids 43
    2.8   Effect of Temperature and Pressure on Surface Tension
           of Liquids 46
           2.8.1   Heat of Liquid Surface Formation and Evaporation 48
    2.9   Interfacial Tension of $Liquid_1$ (Oil)–$Liquid_2$ (Water) 51
           2.9.1   Measurement of IFT between Two Immiscible
                     Liquids 52

Chapter 3 Surface Active and Fracture-Forming Substances (Soaps and
                  Detergents, etc.) 55

    3.1   Introduction 55
    3.2   Surface Tension of Aqueous Solutions (General Remarks) 58

3.2.1　Aqueous Solutions of Surface-Active Substances
(SAS) (Amphiphiles) 60
3.2.2　Solubility Characteristics of Surfactants in Water
(Dependence on Temperature) 62
3.2.2.1　Ionic Surfactants 62
3.2.2.2　Nonionic Surfactants 64
3.3　Micelle Formation of Surfactants (in Aqueous Media)........... 65
3.4　Gibbs Adsorption Equation in Solutions 72
3.4.1　Kinetic Aspects of Surface Tension of Detergent
Aqueous Solutions.. 81
3.5　Solubilization (of Organic Water-Insoluble Molecules) in
Micelles 83

Chapter 4　Surface Chemistry of Solid Surfaces: Adsorption–Desorption
Characteristics.. 87

4.1　Introduction 87
4.2　Wetting Properties of Solid Surfaces 89
4.2.1　Hydraulic Fracture Fluid Injection and
Wettability of Shales 92
4.2.1.1　Hydraulic Fracturing Fluid (Water
Phase) and Reservoir 92
4.3　Surface Tension (γ_{SOLID}) of Solids 94
4.4　Contact Angle (θ) of Liquids on Solid Surfaces..................... 94
4.5　Measurements of Contact Angles at Liquid–Solid
Interfaces.. 95
4.6　Theory of Adhesives and Adhesion.. 97
4.7　Adsorption/Desorption (of Gases and Solutes from
Solutions) on Solid Surfaces (Shale Gas Reservoirs) 98
4.7.1　Gas Adsorption on Solid Measurement Methods 105
4.7.1.1　Gas Volumetric Change Methods of
Adsorption on Solids 105
4.7.1.2　Gravimetric Gas Adsorption Methods 106
4.7.1.3　Langmuir Gas Adsorption 106
4.7.2　Various Gas Adsorption Analyses 107
4.7.3　Adsorption of Solutes from Solution on Solid
Surfaces 109
4.7.4　Solid Surface Area (Area/Gram) Determination 110
4.8　Surface Phenomena in Solid-Adsorption and Fracture
Process (Basics of Fracture Formation) 113
4.9　Heats of Adsorption (Different Substances) on Solid
Surfaces 113
4.10　Solid Surface Roughness (Degree of Surface Roughness).... 115
4.11　Friction (Between $Solid_1$–$Solid_2$)... 115
4.12　Phenomena of Flotation (of Solid Particles To Liquid
Surface) (Wastewater—Hydraulic Fracking) 115

Chapter 5 Solid Surface Characteristics: Wetting, Adsorption, and Related
Processes 119

    5.1   Introduction 119
    5.2   Oil and Gas Recovery (Conventional Reservoirs) and
           Surface Forces 120
           5.2.1   Oil Spills and Clean-Up Process on Oceans 122
           5.2.2   Different States of Oil Spill on Ocean (or Lakes)
                    Surface 122
    5.3   Surface Chemistry of Detergency 124
    5.4   Evaporation Rates of Liquid Drops 126
    5.5   Adhesion ($Solid_1$–$Solid_2$) Phenomena 127

Chapter 6 Colloidal Systems: Wastewater Treatment: Hydraulic Fracking
Technology 131

    6.1   Introduction 131
    6.2   Colloids Stability Theory Derjaguin–Landau–Verwey–
           Overbeek (DLVO) Theory: Silica (Proppant) Suspension
           in Hydraulic Fracking 134
           6.2.1   Charged Colloids (Electrical Charge Distribution
                    at Interfaces) 137
           6.2.2   Electrokinetic Processes of Charged Particles in
                    Liquids 141
    6.3   Stability of Lyophobic Suspensions 142
           6.3.1   Kinetics of Coagulation of Colloids 145
           6.3.2   Flocculation and Coagulation of Colloidal
                    Suspension 146
    6.4   Wastewater Treatment and Control (Zeta Potential) 147

Chapter 7 Foams and Bubbles: Formation, Stability and Application 151

    7.1   Introduction 151
    7.2   Bubbles and Foams 151
           7.2.1   Application of Foams and Bubbles in Technology 152
    7.3   Foams (Thin Liquid Films) 153
           7.3.1   Foam Stability 155
           7.3.2   Foam Formation and Surface Viscosity 158
           7.3.3   Antifoaming Agents 159
    7.4   Wastewater Purification (Bubble Foam Method) 159
           7.4.1   Froth Flotation (An Application of Foam) and
                    Bubble Foam Purification Methods 160
    7.5   Applications of Scanning Probe Microscopes (STM,
           AFM, FFM) to Surface and Colloid Chemistry 161
           7.5.1   Measurement of Attractive and Repulsive Forces
                    (By AFM) 164
                    7.5.1.1   Shale Rock and Other Solid Surfaces 164

Chapter 8   Emulsions and Microemulsions: Oil and Water Mixtures 167

8.1   Introduction 167
      8.1.1   Emulsions and Hydraulic Fracking 168
8.2   Structure of Emulsions 168
      8.2.1   Oil–Water Emulsions 169
      8.2.2   HLB Values of Emulsifiers 170
      8.2.3   Methods of Emulsion Formation............................ 173
8.3   Emulsion Stability and Analyses....................................... 175
      8.3.1   Electrical (Charge) Emulsion Stability.................... 176
      8.3.2   Creaming or Flocculation of Drops 177
8.4   Orientation of Amphiphile Molecules at Oil–Water
      Interfaces 178
8.5   Microemulsions (Oil–Water Systems) 178
      8.5.1   Microemulsion Detergent.................................... 180
      8.5.2   Microemulsion Technology for Oil Reservoirs........ 181

References 183

Appendix I: Geochemistry of Shale Gas Reservoirs (Shale and Energy) 191

Appendix II: Hydraulic Fracking Fluids (Surface Chemistry) 201

**Appendix III: Effect of Temperature and Pressure on Surface Tension of
Liquids (Corresponding States Theory)** 205

**Appendix IV: Solubility of Organic Molecules in Water: A Surface
Tension—Cavity Model System (Structure of Water and Gas Hydrates)** 209

Appendix V: Gas Adsorption–Desorption on Solid Surfaces 213

Appendix VI: Common Physical Fundamental Constants 217

Index 219

Author

K. S. Birdi received his BSc (with honors) from Delhi University, Delhi, India, in 1952 and later majored in chemistry at the University of California at Berkeley, graduating in 1957. After graduation, he joined Standard Oil of California in Richmond, CA. In 1959, he moved to Copenhagen, where he joined Lever Bros. as chief chemist in the Development Laboratory. During this period, he became interested in surface chemistry and joined the Institute of Physical Chemistry, Danish Technical University, Lyngby, Denmark as assistant professor in 1966. Initially he researched aspects of surface science (e.g., detergents, micelle formation, adsorption; lipid monolayers (self-assembly structures), and biophysics). Later, during the early exploration and discovery stages of oil and gas in the North Sea, he collaborated with the Danish National Research Science Foundation program, and other research institutes in Copenhagen, to investigate surface science phenomena in oil recovery. Research grants were awarded by European Union research projects (related to enhanced oil recovery). The projects involved extensive visits to other universities and collaboration with visiting scientists in Copenhagen. He was appointed research professor at the Nordic Science Foundation in 1985 and was appointed professor of physical chemistry at the School of Pharmacy, Copenhagen, in 1990 (retired in 1999). Throughout his career, he has remained involved with industrial contract research programs to retain awareness of real world issues, and to inform research planning.

He has been a consultant to various national and international industries, a member of chemical societies, and a member of organizing committees of national and international meetings related to surface science, and was an advisory member of the journal *Langmuir* from 1985 to 1987.

He has been an advisor for advanced student and PhD projects. He is the author of over 100 papers and articles.

To describe research observations and data he realized that it was essential to publish on the subject. His first book on surface science, *Adsorption and the Gibbs Surface Excess*, Chattoraj, D.K. and Birdi, K.S., Plenum Press, New York was published in 1984. Further publications include *Lipid and Biopolymer Monolayers at Liquid Interfaces*, K.S. Birdi, Plenum Press, New York, 1989; *Fractals, in Chemistry, Geochemistry and Biophysics*, K.S. Birdi, Plenum Press, New York, 1994; *Handbook of Surface & Colloid Chemistry*, K.S. Birdi (Editor) (first edition, 1997; second edition, 2003; third edition, 2009; fourth edition, 2016; CD-ROM 1999), CRC Press, Boca Raton, FL; *Self-Assembly Monolayer*, Plenum Press, New York, 1999; *Scanning Probe Microscopes*, CRC Press, Boca Raton, FL, 2002; *Surface & Colloid Chemistry*, CRC Press, Boca Raton, FL, 2010 (translated to Kazakh, Almaty, Kazakhstan, 2013); *Introduction to Electrical Interfacial Phenomena*, K.S. Birdi (Editor), CRC Press, Boca Raton, FL, 2010; and *Surface Chemistry Essentials*, CRC Press, Boca Raton, FL, 2014). Surface chemistry is his major area of research interest.

1 Surface Chemistry and Geochemistry of Hydraulic Fracturing

1.1 INTRODUCTION

Man has been using fire as an energy source for almost half a million years. Mankind's need for energy (fire, electricity, combustion engines, heating and air-conditioning, and so on: all kinds of everyday energy needs) has been increasing at a rate of about 2% per year (in proportion to the world population increase) over the past decades. Modern mankind (ca. 7 billion people) is thus totally dependent on energy (as related to food, transport, housing and building, medicine, clothing, drinking water, and protection against natural catastrophes (floods, earthquakes, storms, etc.) to sustain human life on earth. For example, one of the most energy-consuming essential products for sustaining life on earth for mankind is *food*. The major sources of energy during the past decades have been

- Wood
- Coal
- Oil
- Gas (methane)
- Hydro-energy
- Atomic energy
- Solar energy
- Wind energy, and so on

At present, oil (about 100 million barrels per day), gas (about 30% of oil equivalent), and coal (about 30% of oil equivalent) are the biggest sources of energy worldwide (Appendix I). The origin of coal (solid), oil (liquid), and gas (mostly methane) has been the subject of extensive research. Chemical analyses have shown that coal, oil, and gas (mostly methane) have been created inside the earth over millions of years from plants, insects, and so on (under high pressure and temperatures) (Burlingame et al., 1965; Levorsen, 1967; Calvin, 1969; Tissot and Welte, 1984; Yen and Chilingarian, 1976; Russell, 1960; Obrien and Slatt, 1990; Jarvie et al., 2007; Singh, 2008; Bhattacharaya and MacEachem, 2009; Slatt, 2011; Zheng, 2011; Zou, 2012; Melikoglu, 2014) (Appendix I).

Furthermore, it is known that there are vast reserves of coal, oil, and gas under the surface of the earth. In this context, it is important to mention that the core of the earth is known to be a region of very high temperature (6000°C) and pressure (Appendix I) as compared with its surface (1 atmosphere pressure; average

1

temperature around 25°C near the equator). This gradient in energy difference means that dynamics exist in the diffusion (migration) flow of fluids and gases. For example, it is reported that methane is present in the inner core of the earth. The flow takes place through fractures and fissures in the earth matrix. In other words, most of the phenomena on the surface of the earth are maintained at much lower temperature and pressure than inside the core of the earth. This also suggests that many fluids/gases (such as oil, lava, and gas) found in the inner core of the earth are at a *higher potential* compared with the surface of earth (in a low temperature and pressure state). This indicates that the natural phenomena on the earth are not static as regards physical and chemical thermodynamics. Hence, these materials (such as gases and fluids [lava, oil]) are able to migrate upward toward the surface of the earth due to the difference in energy through natural cracks and fractures (i.e., fluid/gas flow through porous rocks). Oil or gas is known to be found in two different kinds of reservoirs (Appendix I) (Figure 1.1):

- Conventional sources
- Nonconventional sources (source rock)

Conventional reservoirs are pockets in which the material (oil/gas) that has migrated from source rock has become trapped in the rock structure. Oil/gas has been produced from these conventional reservoirs for almost a century. The conventional reservoirs exhibit physico-chemical characteristics that are different from those of the source rocks (nonconventional) (i.e., from where the oil/gas material has migrated). As regards the origin of oil/gas, it is suggested that this has been generated from plants, animals, and so on over millions of years and is found to be trapped within the source rock (such as shale reservoirs). In a different context, one finds large reserves of methane in the form of *hydrates* in ice, in many parts of the globe (Kvenvolden, 1995; Aman, 2016; Bozak and Garcia, 1976) (Appendix I). The

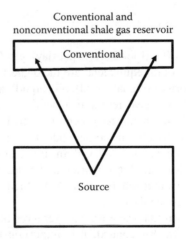

FIGURE 1.1 Oil/gas reservoirs are defined as (a) conventional or (b) nonconventional (source rock).

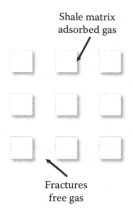

Shale matrix
adsorbed gas

Fractures
free gas

FIGURE 1.2 Shale gas reservoir (shale matrix–adsorbed gas–fractures [free gas]).

supply of oil and gas from conventional reservoirs has been decreasing during the past decades. This has resulted in an urgent need to explore new sources of energy. Both oil and gas have been found in some parts of the earth where the shale (oil/gas) reserves are known to be of very large quantities (e.g., oil reserves of over 10 *trillion* barrels!). Further, during the past decade, gas has been recovered from shale reservoirs (*nonconventional*) in large quantities (mostly in the United States and Canada). This technology is being extensively analyzed in the current literature, and there are some aspects that require more detailed analyses, since the physico-chemical phenomena in such processes are complex. In all kinds of phenomena in which one phase (liquid or gas) moves through another medium (such as porous rocks), the role of *surface forces* becomes important. In the current literature, one finds that the surface chemistry of reservoirs has been investigated at different levels (Bozak and Garcia, 1976; Borysenko et al., 2008; Scheider et al., 2011; Zou, 2012; Josh et al., 2012; Deghanpour et al., 2013; Striolo et al., 2012; Engelder et al., 2014; Mirchi et al., 2015; Birdi, 2016; Scesi and Gattinoni, 2009). This system can be described basically as being composed of *macroscopic* and *microscopic* phases (Figure 1.2).

Shale gas reservoir structure: macroscopic structure-microscopic structure

The macroscopic technology is related to the design of pumps, pipes and tubing, transport, pressure regulation, and so on. The microscopic analyses are related to the essential principles of fluid and gas flow at the production well. This analysis is generally based on laboratory-scale experiments and data, using samples of reservoir rocks. The recovery process from shale reservoirs has been found to be different from those from conventional reservoirs. This is obviously as one would expect. One of the main differences arises from the use of *horizontal* drilling, which allows greater recovery than vertical drilling (Appendix I). Further, shale gas recovery is a multistep process:

- Step I: High-pressure water injection (with suitable additives and creation and stabilization of fractures)
- Step II: Gas recovery (desorption process and diffusion through fractures)

In Step I, the process is related to *surface forces* between water and shale. The initiation of the fracture process is where the molecules at the surface of the rock are involved. This means that surface forces determine the fracture formation. Further, the fluid flow will be described by the classical *flow of liquids* through porous material. The gas recovery (Step II) (i.e., gas desorption) is described from the *solid–gas* interaction theories of surface chemistry (Chapter 4). The first step is mainly the liquid flow through porous media. This is known to be related to *capillary forces* (Chapter 2). The second step is found to be the flow of gas (methane) through very narrow pores (Howard, 1970; Tucker, 1988; Civan, 2010; Javadpour, 2009; Allan and Mavko, 2013; Engelder et al., 2014; Yew and Weng, 2014). It is also suggested that most of the gas is in an adsorbed state (Hill and Nelson, 2000; Shabro, 2013; Ozkan et al., 2010). Experiments have shown that this is a reasonable assumption. It is reported that gas (mostly methane) is self-generating in shale, and that free gas and adsorbed gas coexist. Methane, as an organic molecule, will also be expected to adsorb to the organic (kerogen) part of the shale (Appendix I). The oil–shale (illite clay) adhesion characteristics have been investigated (Bihl and Brady, 2013). The impact of hydraulic fracturing and the degree of flow-back have also been studied.

The *adsorption–desorption* surface chemistry principles of gases on solid surfaces have been investigated in the literature (Adam, 1930; Chattoraj and Birdi, 1984; Adamson and Gast, 1997; Holmberg, 2002; Matijevic, 1969–1976; Somasundaran, 2015; Birdi, 2016) (Chapter 4). It is also estimated that 20%–80% of the total gas in place is present in the adsorbed state. The complex description of the gas shale reservoir is delineated in Figure 1.3. It is thus obvious that this technology requires a long-term production research and development approach. The surface area over which gas is adsorbed is also very extensive. *Surface diffusion* is the important step in the flow and recovery of gas (Bissonnette et al., 2015). If the pores are >50 nm (macropores), then the collision frequency between gas molecules will be expected to be greater than the frequency of collisions between gas and the solid surface. In the case where the gas molecule free path length is larger than the pore diameter, the frequency of collision between gas molecules dominates the process (the so-called Knudsen diffusion domain [Appendix II]). Surface diffusion dominates in micropores (<2 nm diameter). Accordingly, the pressure, the temperature, the solid surface, and the interaction parameters between the gas and the solid surface determine surface diffusion.

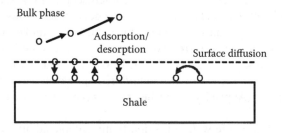

FIGURE 1.3 Gas-recovery process in shale reservoir: (a) adsorption–desorption; (b) diffusion in pores and surface diffusion.

This description of a shale gas reservoir is the most plausible in the current literature. The science of surface chemistry has been applied in various technologies (such as geology, geophysics, geochemistry, hydrology, reservoir engineering, petroleum exploration, biochemistry, paper and ink, and cleaning and polishing) (Birdi, 1997, 2003, 2014, 2016).

In any system where one material (oil, gas, or water) is flowing through given surroundings (porous material such as rock, etc.), this requires knowledge of the *interfacial chemistry*. An interface is the contact area between two different phases (i.e., surfaces such as oil–rock, gas–rock, water–rock, and oil–water). The *surface chemistry* of such systems is known to be the determining factor. In this book, the essential principles of surface chemistry in gas shale reservoirs will be delineated. Especially, the role of hydraulic fracking will be delineated. High-pressure injection of water (with suitable *surface active fracture substances* (SAFS)) is used to create fractures in the shale rock. This system creates *new* solid surfaces (i.e., a fracture). Thus, *fracture formation* requires the understanding of surface forces present in rocks. The fracture (or cracking) phenomena of solids (under stress) have been investigated by surface chemistry principles (Rehbinder and Schukin, 1972; Shipilov et al., 2008; Malkin, 2012; Adamson and Gast, 1997; Birdi, 2014, 2016) (Chapter 4). Further, some basic aspects of surface and colloid chemistry in gas and oil reservoirs will be delineated.

To explain these systems in more detail, it is important to consider the structure of matter. The matter which the universe is made of has been generally described by classic physics and chemistry. All natural phenomena are related to reactions and changes, which are dependent on the structures of the matter involved (Figure 1.4):

- Solids
- Liquids
- Gases

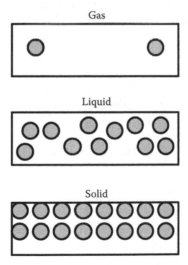

FIGURE 1.4 Solid–liquid–gas.

However, in many industrial (chemical industry and technology) and natural biological phenomena, one finds that some processes require a more detailed definition of matter. This is generally the case when two different phases meeting (e.g., liquid–air, solid–air, liquid–solid, liquid [A], liquid [B], solid [A], and solid [B]) are involved. The combinations of phases are described in the following subsections.

Solid Phase—Liquid Phase—Gas Phase

The molecular structure in these phases differs, and thus, the phenomena related to the individual phases will need information on each particular state of matter. For instance, in the case of a shale gas reservoir, one may depict this system as

Different Surface Phases in a Shale Gas Reservoir: Solid Phase (Shale)— Liquid Phase (Fracking Fluid)—Gas Phase (Methane Gas)

There are two distinct aspects, different in their characteristics (Figure 1.5), that are relevant to surface chemistry principles. These surface chemistry aspects have been recognized to be helpful in understanding the fundamental forces involved in gas recovery. Further, the different stages of the process are analyzed with respect to the surfaces involved; for example,

- Shale surface (solid surface)
- Hydraulic fracture formation (solid–liquid interface)

Gas (mostly methane, CH_4) production from shale reservoirs is an important example where the above relation is of basic interest. It is known that in all transport (flow) systems (such as oil in the reservoir or gas in the shale), the bottleneck is the surfaces (interfaces) involved. The shale rock is known to be very compact, with very low permeability (Appendix I). Further, the matrix pores in the shale are found to consist of different kinds:

- Organic phase
- Inorganic phase

Fluid injection

Fracture formation

Gas diffusion

FIGURE 1.5 Flow of water phase through the porous rock (fracture formation)—gas diffusion from the shale.

This requires the creation of fractures through which the gas (adsorbed on the rock) can be recovered. In the oil/gas industry, the fracking technique has been used for many decades (Appendix II). A *fracking water solution* (hydraulic) is injected into the reservoir under very high pressure (Cahoy et al., 2013; Engelder et al., 2014). The water phase gives rise to fractures of different sizes. Furthermore, the wetting properties of rocks have been investigated by surface chemistry principles (Chapter 4). These aspects are important for the water injection (hydraulic fracture) technology (Borysenko et al., 2009; Engelder et al., 2014). Especially, the significance of the wetting as determined by the hydrophilic–hydrophobic characteristics of shale rocks has been investigated.

1.2 FORMATION OF FRACTURES IN SHALE RESERVOIRS AND SURFACE FORCES

Gas shale rocks exhibit very low permeability. It has therefore been found that one needs to create fractures and fissures in the gas-bearing bedrock for enhanced gas recovery. The process used is called *hydraulic fracturing*. This consists of using fracturing fluids (water with the necessary additives) to create fractures by the application of high pressure. In general, the hydraulic fracture process is composed of the following steps:

- High-pressure fluid injection
- Creation of fractures
- Gas (mostly methane) desorption and diffusion (through fractures) to the surface of the earth

It is thus obvious that in this process, various surfaces (interfaces) are involved:

- Water fracture solution (liquid phase) and shale rock (solid phase)
- Fracture formation (initial step at the surface of the rock)
- Gas recovery (gas phase) and shale rock (solid phase)

Various interfaces are involved in these phenomena, which indicates that primarily, surface forces are involved. The fracture, for example, is known to initiate at the surface (i.e., surface forces interacting between the molecules) of the rock. The gas in the shale (source rock) is present at higher potential than at the borehole, and hence will eventually diffuse (through fractures) to the surface of the earth (over a geological timescale of thousands of years!).

In the reservoirs, fracturing is created when fluid is pumped at a faster rate than it can be absorbed by the rock formation. The injection of high-pressure water solution is found to create multiple fractures due to mechanical stress. This is a process whereby one creates (breaks) two new solid surfaces (related to the surface forces). However, there have also been reports of fracking by using other fluids (emulsions, foams, etc.). The fractures are stabilized by the addition of 5%–10% small silica particles (proppants) (or other solid particles of similar properties) to the fracking solution. As well as high-pressure water (95%–90%), the fracking solution also contains necessary additives (lower than 2%) (Appendix II), which are based on the following physico-chemical properties and functions:

- Silica particles (in suspension) to stabilize the fractures
- Polymers (high viscosity)
- Gelling agents
- Surface-active substances (SAS)
- SAFS (fracture formation)
- Foaming agents
- Other additives, such as pH control, biocides, corrosion inhibitors

It is obvious that the fracking process can be expected to be complicated in the case of shale matrix. The application of SAFS additives has been reported in the literature (and patents) in similar kinds of phenomena (such as cracks and fracture formation in solids). The fracture process basically means that solid material in the rock (or a metal) is separated into two (solid) new surfaces with a liquid (fluid, emulsion, etc.) in between (Figure 1.6).

After the fracture is created, the gas is desorbed (from the surface of shale rock) and diffuses through the fractures (pores). Gas desorption (equilibrium and rate) is dependent on the equilibrium between the adsorbed and desorbed states of the system. The thermodynamics of this surface process is being investigated in the current literature. The fracture formation will thus be dependent on both the properties of the rock (surface forces of the solid) and the liquid injected (generally water, plus any additives such as alcohol or SAFS). SAFS are those additives that facilitate fracture formation in solids. This subject has been investigated in the literature (Aderibigbe, 2012; Dunning et al., 1980; Santos, 2008; Boschee, 2012; Engelder et al., 2014; Ma and Holditch, 2016). It has been known for many decades that solids (crystals, rocks,

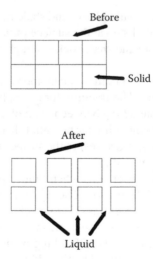

FIGURE 1.6 Schematic of fracture formation by the injection of liquid solution (before and after hydraulic injection).

or metals) exhibit a crack-propagation process (under mechanical stress) that initiates at the surface (molecular) region (Chapter 4) and spreads toward the bulk phase. On a molecular level, this implies that the cracks are initiated at the surface molecules of the solid material. It has also been reported that specific additives (surface-active fracture substances: SAFS) to water can induce the fracture process. Many investigations have been carried out on pure rock crystals and pure metals. The crack propagation is suggested to initiate from the surface layers of molecules (Figure 1.7):

- Solid: surface molecules
- Crack propagation
- Bulk solid phase

For example, analogous fracture (or crack) formation in different systems has been known for many decades. These fracture studies were based on different systems that one finds in everyday life (Figure 1.7). Any solid breaks under suitable applied pressure. However, if one *scratches* the surface of glass (with a diamond cutter), then it will break precisely (on application of pressure) at the line of scratch (surface phenomena). A pure metal breaks at the line of defect after another metal has been used to scratch its surface (such as zinc and gallium) (Rehbinder and Schukin, 1972). This suggests that for fracture (crack propagation) to initiate, one has to change the interaction energy between the surface molecules (i.e., surface forces) (Figure 1.7). It was found that in some rocks, the surface charge (i.e., zeta potential) of the fluid environment is important for the initial step in fracture formation. Further, it has been reported that in general, fracture formation is related to the surface properties of the added solute. In this context, the surface adsorption property of the SAFS additives is important. The interfacial adsorption of any solute in water has been described in

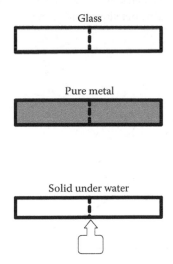

FIGURE 1.7 Idealized fracture formation in a solid: (a) glass with a scratch; (b) pure metal after a swipe over with another metal; (c) fracture of a solid while covered with a water solution.

the literature by the general Gibbs adsorption theory applied to all kinds of adsorption at interfaces, e.g., liquid–gas, liquid–liquid, liquid–solid, solid–gas (Adamson and Gast, 1997; Chattoraj and Birdi, 1984; Birdi, 2016; Somasundaran, 2015; Fathi and Yucel, 2009).

The flow of gas in any porous solid matrix is related to the interfacial forces, that is, gas–solid. The movement of gas in shale (in the organic phase, i.e., kerogen), means that gas molecules are found in the following phases (Scheidegger, 1957; Letham, 2011; Shabro et al., 2011a, 2011b, 2012; Birdi, 1997, 2016; Fengpeng et al., 2014; Kumar, 2005; Rao, 2012):

- Gas diffusion (i.e., movement of gas through the fractures)
- Adsorption/desorption energy (the adsorbed gas, mainly on the organic part of the shale, has to desorb to escape to the surface of the well)

Some investigations carried out on shale core samples indicate that adsorption–desorption of methane follows Langmuir adsorption laws (Bumb and Mckee, 1988; Kumar, 2012) (Chapter 4). Further, current production analyses indicate that the gas in shale reservoirs exists in distinct phases (Fathi and Yucel, 2009):

- Free gas in the fractures
- Adsorbed gas on the shale
- Dissolved gas in brine water (very low)

The rate of recovery will be primarily dependent on the potential difference between the free and adsorbed gas phases. The rate of gas recovery has been found to be different for different shale reservoirs (Figure 1.8). This indicates that the gas is adsorbed on shale in different states. It is also observed that the gas production from a shale reservoir is fast initially, but slows with time.

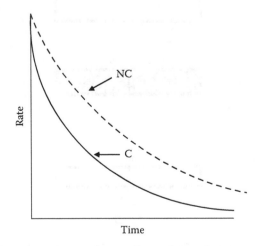

FIGURE 1.8 Rate of gas production from conventional (C) and nonconventional (N) shale reservoirs.

The rate of production is primarily related to the adsorption–desorption energy of the gas molecules on the shale. The production rates have indicated that the primary gas produced is from the free gas, while the secondary production (at a slower rate) is related to the adsorbed gas (Oligney and Economides, 2002; Shabro et al., 2009, 2011a, 2011b, 2012; Donaldson et al., 2013; Yew and Weng, 2014). The surface chemistry of such systems can be analyzed at the microscopic level. If one observes a container of liquid (such as water), one notices that liquid and gas (air) meet at the surface. However, if one takes a molecular snapshot of the system, one finds from experiments that the molecules that are situated at the interfaces (e.g., gas–liquid, gas–solid, liquid–solid, $liquid_1$–$liquid_2$, $solid_1$–$solid_2$) behave differently from those in the bulk phase (Adam,1930; Aveyard and Hayden, 1973; Bancroft, 1932; Partington, 1951; Chattoraj and Birdi, 1984; Davies and Rideal, 1963; Defay et al., 1966; Gaines, 1966; Harkins, 1952; Holmberg, 2002; Matijevic, 1969–1976; Fendler and Fendler, 1975; Adamson and Gast, 1997; Auroux, 2013; Birdi, 1989, 1997, 1999, 2003, 2009, 2016; Somasundaran, 2006, 2015). Typical examples are

- Surfaces of oceans, rivers, and lakes (liquid–air interface)
- Road surface (solid–air or solid–car tire)
- Lung surface
- Washing and cleaning surfaces
- Emulsions (cosmetics and pharmaceutical products)
- Oil and gas reservoirs (conventional and nonconventional)
- Diverse industries (paper, milk products)

For instance, reactions taking place at the surface of oceans will be expected to be different from those observed inside the seawater. Further, in some instances, such as oil spills, one can easily realize the importance of the role of the surface of oceans. It is also well known that the molecules situated near or at an interface (i.e., liquid–gas) will be interacting differently with respect to each other than the molecules in the bulk phase (Figure 1.9a and b). The intramolecular forces acting will thus be different in these two cases. In other words, all processes occurring near any interface will be dependent on these molecular orientations and interactions. Furthermore, it has been pointed out that, for a dense fluid, the repulsive forces dominate the fluid structure and are of primary importance. The main effect of the repulsive forces is to provide a uniform background potential in which the molecules move as hard spheres. The molecules at the interface will be under an asymmetrical force field, which gives rise to the so-called surface tension or interfacial tension (Figure 1.9) (Chattoraj and Birdi, 1984; Birdi, 1989, 1997, 1999, 2003, 2016; Adamson and Gast, 1997; Somasundaran, 2015).

This leads to the adhesion forces between liquids and solids (Chapter 5), which are a major application area of surface and colloid science.

The resultant force on molecules will vary with time because of the movement of the molecules in the liquid state. The molecules at the surface will be under the influence of forces that are mostly directed downward into the bulk phase. The nearer the molecule is to the surface, the greater the magnitude of the force due to *asymmetry*. The region of asymmetry plays a very important role (near all kinds of surfaces).

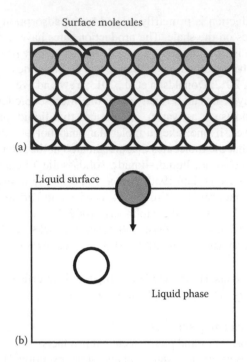

FIGURE 1.9 (a) Surface molecules (shaded) (b) intermolecular forces around a molecule in the bulk liquid (dark) and around a molecule in the surface (light).

Thus, when the surface area of a liquid is increased, some molecules must move from the interior of the continuous phase to the interface. The surface tension of a liquid is the force acting normal to the surface per unit length of the interface, thus tending to decrease the surface area. The molecules in the liquid phase are surrounded by neighboring molecules, and these interact with each other in a symmetrical way. In the gas phase, where the density is 1000 times lower than in the liquid phase, the interactions between molecules are very weak compared with those in the dense liquid phase. Thus, when one crosses from the liquid phase to the gas phase, there is a 1000-fold change in density. This means that in the liquid phase a molecule occupies a volume that is 1000 times smaller than when it is in the gas phase.

Surface tension is the differential change of free energy with change of surface area. An increase in surface area requires that molecules from the bulk phase are brought to the surface phase. The same is valid when there are two fluids or a solid and a liquid; this is usually designated *interfacial tension*. A molecule of a liquid attracts the molecules that surround it, and in turn, it is attracted by them (Figure 1.9). For the molecules inside a liquid, the resultant of all these forces is neutral, and all of them are in equilibrium and reacting with each other. When these molecules are on the surface, they are attracted by the molecules below and to the side, but not toward the outside (i.e., the gas phase). The resultant is a force directed inside the liquid. In its turn, the cohesion among the molecules supplies a force tangential to the surface. Hence, a fluid surface behaves like an elastic membrane that wraps and compresses the liquid below. The surface tension expresses the force with which the surface molecules attract each other.

In fact, it has been found that a metal needle (heavier than water) can be made to float on the surface of water (if it is carefully placed on the surface). The surface of a liquid can thus be regarded as the plane of potential energy. It may be assumed that the surface of a liquid behaves as a *membrane* (on a molecular scale), which stretches across the liquid and needs to be broken to penetrate it. The reason a heavy object floats on water is that for it to sink, it must overcome the surface forces. This clearly shows that at any liquid surface, there exists a tension (surface tension), which needs to be broken when any contact is made between the liquid surface and the material (here, the steel needle). A liquid can form three types of interfaces:

1. Liquid and vapor or gas (e.g., ocean surface and air)
2. Liquid$_1$ and liquid$_2$ immiscible (water–oil, emulsion)
3. Liquid and solid interface (water drop resting on a solid, wetting, cleaning of surfaces, adhesion)

As regards solid surfaces, these can similarly exhibit additional characteristics:

1. Solid–solid (fracture formation, earthquake)
2. Solid$_1$–solid$_2$ (cement, adhesives)

Furthermore, a fracture (__//__) is created when a solid material is broken into two separate entities (plates):

Solid fracture to form two *new* surfaces:

Start =====

After __//__//__
              _//___//__.............Fracture

Solid and gas before a fracture:

Start.....===Gas===Gas====

Solid and gas after fracture:
         ____//_____//_____
         ____//_Gas____//_____.........Fracture
         _Gas//_____
         ____//____Gas_//_____

The adsorbed gas is desorbed after fracture formation and pressure drop. Each fracture formation means that essentially, two new solid surfaces are created by the hydraulic (mechanical or other means) process (Figure 1.10). In other words, energy (*surface energy*) is needed to create a definite *fracture surface area*. The mechanical

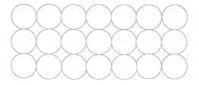

Surface fracture

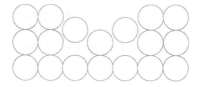

FIGURE 1.10   Fracture formation at its initial state (surface defect).

energy input is proportional to the (fracture) surface energy needed to create the fracture surface area:

$$\text{Fracture surface energy} = \text{surface tension} \left(\text{of the solid}\right)$$
$$\times \text{surface area of the fracture} \qquad (1.1)$$

In these different processes, the surface energy (solid surface tension) involved has been investigated based on surface chemistry principles (Rehbinder and Schukin, 1972; Latanision and Pickens, 1983; Adamson and Gast, 1997; Birdi, 2016; Somasundaran, 2015). These investigations have shown that some additives that adsorb at the water surface (SAFS) (Chapter 3) induce the fracture process. From surface chemistry principles, this means that the fracture formation energy is related to the surface force (i.e., surface tension). The latter quantity can be changed by changing the necessary physical parameters of the system (such as pH, ionic strength, or additives that are surface active). The fracture in any solid is initiated at the surface:

$$\text{Fracture} \rightarrow \text{Initial formation begins at the surface}$$

The surface molecules have to be separated from each other, and this leads to the fracture (Javadpour et al., 2007). Some general fracture systems are

1. Glass cracking: mechanical process (after scratch)
2. Metal cracking: surface molecular
3. Rock under water: effect of surface tension or surface charges
4. Shale rock (or similar): complex process (combination of II and III)

These different processes have been described in the literature (Chapter 4). In the earth, one finds fractures created by natural phenomena (such as earthquakes).

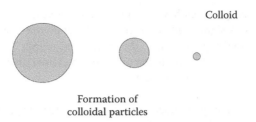

Colloid

Formation of
colloidal particles

FIGURE 1.11 Formation of fine particles (Chapter 4) (schematic: less than micrometer size).

It is through these fractures that the oil or gas has migrated into the conventional reservoirs from the source rocks. In a different context, the forces needed for this fracture formation can be compared to those when the solid is crushed and surface area increases per unit gram (Figure 1.11). For example, finely divided talcum powder has a surface area of 10 m^2/g. Active charcoal exhibits surface areas corresponding to over 1000 m^2/g. This is an appreciable figure, and its consequences will be delineated later. Qualitatively, one must notice that work has to be put into the system when one increases the surface area per gram (weight) (for liquids, solids, or any other interface). This is most important in the cement industry. Creating finer particles of cement requires the input, and accordingly the cost and production, of considerable energy. It is also known that specific additives have been used that reduce the energy needed to create fine particles in the cement industry (and many other processes, such as in the drilling industry).

The surface chemistry of small particles is an important part of everyday life (such as dust, talcum powder, sand, raindrops, and emissions). The designation *colloid* is used for particles that are of such small dimensions that they cannot pass through a membrane with a pore size ca. 10^{-6} m (=μm) (Birdi, 1997, 2016). The nature and relevance of colloids have been a major research topic over many decades (Birdi, 2003a). Colloids are an important class of materials, intermediate between bulk and molecularly dispersed systems. The colloid particles may be spherical or elliptical, but in some cases, one dimension can be much larger than the other two (as in a needle-like shape). The size of particles also determines whether one can see them with the naked eye. Colloids are not visible to the naked eye or under an ordinary optical microscope. Scattering of light can be used to see such colloidal particles (such as dust particles). The size of colloidal particles may range from 10^{-4} to 10^{-7} cm. The units used are as follows:

- 1 μm = 10^{-6} m
- 1 nm = 10^{-9} m
- 1 Å (Angstrom) = 10^{-8} cm = 0.1 nm = 10^{-10} m

Nanosize (in the nanometer range) particles are currently of much interest in different applied science systems (*nano* is derived from Greek and means *dwarf*). Nanotechnology has been strongly boosted by the last decade of innovation, as reported by the surface and colloid literature. Since colloidal systems consist of two or more phases and components, the interfacial area to volume ratio becomes very significant. Colloidal particles have a high surface area to volume ratio compared with bulk materials. A significant proportion of the colloidal molecules lie within, or close to, the interfacial region. Hence, the interfacial region has significant control over the properties of colloids. To understand why colloidal dispersions can be either stable or unstable, one needs to consider

1. The effect of the large surface area to volume ratio (e.g., 1000 m^2 surface area per gram of solid (active charcoal, etc.))
2. The forces operating between the colloidal particles (ratio between particle size and distance of separation)

There are some very special characteristics that must be considered as regards colloidal particle behavior: size and shape, surface area, and surface charge density.

In the *fracking* process, generally, silica particles are dispersed in water. This process is the colloidal surface chemistry aspect. The application of silica particles is to stabilize the fractures for gas desorption. The stability of fracking silica solution is based on colloidal theory.

It is thus found that some terms need to be defined at this stage. The definitions generally employed are as follows. *Surface* is the term used when one considers the dividing phase between

* Gas–liquid
* Gas–solid

Interface is the term used when one considers the dividing phase between

* Solid–liquid: Colloids
* Liquid$_1$–liquid$_2$: Oil–water, emulsion
* Solid$_1$–solid$_2$: Adhesion (glue, cement), fracture/crack formation, drilling

In other words, the surface tension may be considered to arise due to a degree of unsaturation of bonds that occurs when a molecule resides at the surface and not in the bulk. The term *surface tension* is used for solid/vapor or liquid/vapor interfaces. The term *interfacial tension* is more generally used for the interface between two liquids (oil–water), two solids, or a liquid and solid. It is, of course, obvious that in a one-component system, the fluid is uniform from the bulk phase to the surface. However, the orientation of the surface molecules will be different from those molecules in the bulk phase. For instance, in the case of water, the orientation of molecules inside the bulk phase will be different from those at the interface. The hydrogen bonding will orient the oxygen atom toward the interface. The question one may ask, then, is how sharply the density changes from that of a fluid to that of gas (a change by a factor of 1000). Is this a transition region a monolayer deep or multilayers deep? This subject has been extensively investigated for almost a century. The most important theoretical analyses have been provided by the Gibbs adsorption theory, which relates the surface property to the change in bulk phase. The Gibbs adsorption theory (Birdi, 1989, 1999, 2003, 2016; Defay et al., 1966; Chattoraj and Birdi, 1984) considers the surface of liquids to be a monolayer. The surface tension of water decreases appreciably on the addition of very small quantities of soaps and detergents. Gibbs adsorption theory relates the change in surface tension to the change in soap concentration. Experiments that analyze the spread monolayers are also based on one molecular layer. The latter data, indeed, conclusively verifies the Gibbs assumption. Detergents and other similar kinds of molecules (soaps, etc.) are found to exhibit self-assembly characteristics (i.e., aggregate-forming systems) (Tanford, 1980; Birdi and Ben-Naim, 1980; Birdi, 1999; Somasundaran, 2015).

1.3 COLLOIDS

Colloids is Greek for *glue-like*. It has been known for centuries that the property of a solid changes when its size is reduced. One finds in everyday life a wide variety of systems consisting of finely divided particles (talcum powder, sand and dust, nano-size particles, and so on) or macromolecules (glue, gelatin, proteins, etc.) (Table 1.1).

Colloidal systems are widespread in their occurrence and have biological and technological significance. For example, in the hydro-fracking fluid, one uses finely divided silica (SiO_2) dispersed in water. The main application of SiO_2 is to keep the fractures stabilized. The surface forces present at SiO_2 and the surrounding phase have been investigated by direct methods (Birdi, 2003). In the latter system, wastewater treatment is also a typical example of surface chemistry principles (Birdi, 1999, 2016). There are three types of colloidal systems (Adamson and Gast, 1997; Birdi, 2003, 2009):

1. In simple colloids, a clear distinction can be made between the disperse phase and the disperse medium, for example, simple emulsions of oil-in-water (o/w) or water-in-oil (w/o).

TABLE 1.1
Typical Colloidal Systems

	Phase	
Dispersed	**Continuous**	**System name**
Liquid	Gas	Aerosol fog, spray
Gas	Liquid	Foam, thin films, froth, fire extinguisher foam
Liquid	Liquid	Emulsion (milk), mayonnaise, butter, oil/water (emulsions)
Solid	Liquid	Sols, AgI, suspension wastewater, cement, metallurgy, paint and ink, hydraulic fracking
Liquid	Solid	Solid emulsion (toothpaste)
Solid	Gas	Solid aerosol (dust), smog
Gas	Solid	Solid foam (expanded polystyrene), insulating foam
Solid	Solid	Solid suspension/solids in plastics
Biocolloids		
Corpuscles	Serum	Blood
Hydroxyapatite	Collagen	Bone, teeth

2. Multiple colloids involve the coexistence of three phases, two of which are finely divided: for example, multiple emulsions (mayonnaise, milk) of water-in-oil-in-water (w/o/w) or oil-in-water-in-oil (o/w/o).
3. Network colloids have two phases forming an interpenetrating network: for example, polymer matrix.

The colloidal (as solids or liquid drops) stability is determined by the free energy (the surface free energy or the interfacial free energy) of the system. The main parameter of interest is the large surface area exposed between the dispersed phase and the continuous phase. Since the colloid particles move about constantly, their dispersion energy is determined by Brownian motion. The energy imparted by collisions with the surrounding molecules at temperature $T = 300°K$ is $3/2 \ k_B T = 3/2 \times 1.38 \times 10^{-23} \times 300 = 10^{-20}$ J (where k_B is the Boltzmann constant). This energy and the intermolecular forces would thus determine the colloidal stability. In the case of colloid systems (particles or droplets), the kinetic energy transferred on collision will thus be $k_B T = 10^{-20}$ J. However, at a given moment, there is a high probability that a particle may have a larger or smaller energy. Further, the probability of total energy being over 10 times $k_B T$ thus becomes very small. The instability will be observed if the ratio of the barrier height to $k_B T$ is around 1–2 units. The idea that two species (solid–solid) should interact with one another, so that their mutual potential energy can be represented by some function of the distance between them, has been described in the literature. Furthermore, colloidal particles frequently adsorb (and even absorb) ions from their dispersing medium (such as in groundwater treatment and purification).

Sorption that is much stronger than what would be expected from dispersion forces is called chemisorption, a process that is of both chemical and physical interest. For example, in shale gas recovery, water and SAFS are injected to induce hydrofracking (SAFS change the surface forces).

1.4   EMULSIONS (AND HYDRAULIC FRACKING FLUIDS)

From common observation, one knows that oil and water do not mix. This suggests that in emulsions, these systems are dependent on the oil–water interface. The liquid$_1$ (oil)–liquid$_2$ (water) interface is found in many systems, most importantly in the world of emulsions (Friberg et al., 2003; Birdi, 2016).

There are two main aspects of emulsions that confront modern technology (Friberg et al., 2003). One is where an oil–water emulsion is needed, such as in cosmetics; the other is where an emulsion is undesirable, such as in wastewater. For example, in the case of the oil gas industry, there are systems where undesirable emulsions are of major concern. For instance, the recovered fracking fluid in back flow may exhibit an oil–water emulsion. The trick in using emulsions is based on the fact that one can apply both water and oil (the latter is insoluble in water) simultaneously. In fracking technology, it has been found that by using emulsions, one can reduce the amount of water in the process (Appendix I).

Further, one can then include other molecules, which may be soluble in either phase (water or oil). This obviously leads to the common observation that one finds thousands of applications of emulsions. It is very important to mention here that actually, nature uses this trick in most of the major biological fluids.

In fact, the state of mixing oil and water is an important example of interfacial behavior at the $liquid_1\_liquid_2$ interface. Emulsions of oil–water systems are useful in many aspects of daily life: milk, foods, paint, oil recovery, pharmaceuticals, and cosmetics. In fact, the colloidal chemistry of milk is the most complicated in a natural product.

If one mixes olive oil with water, on shaking:

- About 1 mm diameter oil drops are formed.
- After a few minutes, the oil drops merge together, and two layers are again formed.

However, if one adds suitable substances, due to surface forces, the olive oil drops formed can be very small (in the micrometer range).

In addition, one finds that these considerations are important in regard to different systems: paints, cements, adhesives, photographic products, water purification, sewage disposal, emulsions, chromatography, oil recovery, paper and print industry, microelectronics, soap and detergents, catalysts, and biology (cells, viruses).

2 Capillary Forces in Fluid Flow in Porous Solids (Shale Formations)

2.1 INTRODUCTION

The flow characteristics of fluids (water, oil, etc.) in porous solids (oil and gas reservoirs [conventional or nonconventional], drinking water, etc.) is important for these systems. For example, in the case of shale gas, one finds that the fracking fluid (hydraulic fracking) moves through the porous matrix and creates or extends (stabilizes) fractures (Figure 2.1). The detailed microscopic analyses of shale reservoirs are found to be explained by the application of surface chemistry principles. Fracture formation (in general) in solids can be delineated as

Solid matrix....Fracture (creation of two new solid surfaces)

The flow of hydraulic fracking fluid (water solution) through shale matrix is the primary step. This is the same as any fluid flow through a porous media (i.e., capillary pressure). In addition, this process has another significant interfacial aspect, which is termed *wetting*. The wetting characteristic is a phenomenon that involves surface forces, which are acting between the water phase and the shale rock (consisting of inorganic and organic material) (Chapter 4). It is known that surface chemistry principles must be applied to understand such systems, since liquid and solid surfaces are involved. Liquids exhibit some unique properties, which are the subject matter of interest in this chapter. Solid surfaces are obviously somewhat different, due to their rigid structure as compared with liquids (Chapter 4).

Hydraulic fracking (see Appendix II) involves flow (injection) of fluids into shale rocks (under high pressures). The (low) porosity of shale is the major factor as regards fracking phenomena. The detailed analyses of such phenomena require analyses of the fluid behavior in a porous matrix (such as shale). It is a common observation that liquids take the shape of the container that surrounds or contains them. However, one also finds that in many cases, there are other subtle properties, which arise at the interface of liquids. Another phenomenon is that when a glass capillary tube is dipped in water, the fluid rises to a given height. It is observed that the narrower the tube, the higher the water rises. The only difference in the case of this observation is that the liquid *curvature* in the different tubes is different. This observation indicates that the mechanical forces at a curved fluid surface are different from those at a more flat surface (Figure 2.2).

Rock

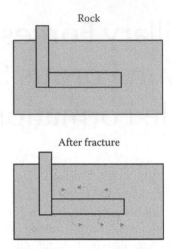

After fracture

FIGURE 2.1   Injection process of high-pressure water phase into shale rock.

Shape of a
liquid surface

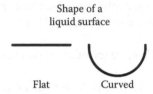

Flat   Curved

FIGURE 2.2   A flat and a curved liquid surface.

This phenomenon is found in a simple system such as a sponge, as well as in more complicated systems such as water or oil flow in reservoirs, water rising in trees, and blood flow in arteries.

The role of liquids and liquid surfaces is important in many everyday natural processes (e.g., oceans, lakes, rivers, raindrops). Therefore, in these systems, one would expect surface forces to be important. Accordingly, one will thus need to study surface tension and its effect on the surface phenomena in these different systems. This means that one needs to consider the structures of molecules in the bulk phase in comparison with those at the surface.

From simple geometrical considerations, one can show that the surface molecules are under a different force field than the molecules in the bulk phase or the gas phase. These forces are called *surface forces*. A liquid surface behaves like a stretched elastic membrane in that it tends to contract. This arises from the observation that when one empties liquid from a beaker, the liquid breaks up into spherical drops. This indicates that drops are being created under some forces that must be present at the surface of the newly formed interface. The spherical shape of the drops is related to the fact that all changes in nature are driven toward lowering energy. As a qualitative description of the surface molecules, one may safely suggest that these are in a state between the bulk liquid and the gas phase. This asymmetry leads to tension in the surface region (a few molecules thick).

These surface forces become even more important when a liquid is in contact with a solid (such as groundwater or an oil reservoir). The flow of liquid (e.g., water or oil) through small pores under the ground is mainly governed by *capillary forces*. It is found that capillary forces play a dominant role in many of the systems that will be described later. Thus, the interaction between a liquid and any solid will form curved surfaces, which, being different from a planar fluid surface, give rise to capillary forces.

2.2   SURFACE FORCES IN LIQUIDS

Physical chemistry studies are based on the molecular interactions in liquids that are responsible for their physico-chemical properties (such as boiling point, melting point, heat of vaporization, and surface tension). This is useful, as one can both describe and relate different properties of matter at a more molecular level (both qualitatively and quantitatively). These ideas are the basis for the quantitative structure activity related (QSAR) (Kubinyi, 1993; Hansch et al., 2002; Cronin, 2004; Birdi, 2003a,b, 2016) analyses of various substances. This approach for analyses and applications is becoming more advanced due to the enormous help available from computer capacity.

All different kinds of forces acting between any two molecules are dependent mainly on the distance between the two molecules, besides other parameters (Birdi, 1997, 2016). To illustrate these aspects, one can consider the following example. One can estimate (semi-quantitatively) the difference in distance between molecules in liquid or gas as follows. For example, in the case of water, the following data is known (a typical example, at room temperature and pressure):

$$\text{Volume per mole liquid water} = V_{liquid} = 18 \text{ mL/mole} \qquad (2.1)$$

$$\text{Volume per mole water in gas state} \left(\text{at STP}\right)\left(V_{gas}\right) = 22 \text{ l/mole} \qquad (2.2)$$

$$\text{Ratio}\left(\frac{V_{gas}}{V_{liquid}}\right) = \text{ca. } 1000$$

Hence, the approximate distance between gas molecules will be ca. 10 times larger than in the liquid phase ($=(1000)^{1/3}$ (from simple geometrical considerations of volume ($=$length^3) and length)). The distance between solid phases is ca. 10% less than in liquids (in general).

In other words, the density of water changes 1000 times as the surface is crossed from the liquid phase to the gas phase (Figure 2.3). Other fluids exhibit almost the same characteristic. This large change near the surface of the liquid means that the surface molecules must be in a different environment than in the liquid phase or the gas phase. The distance between gas molecules is approximately 10 times larger than in a liquid. Hence, the forces between gas molecules are much weaker than in the case of the liquid phase (all forces increase when distances between molecules decrease). All interaction forces between molecules (solid phase, liquid phase, and

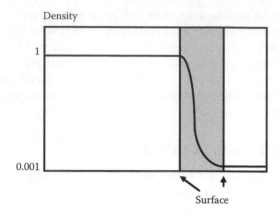

FIGURE 2.3   Change of density of a fluid (water) near the surface.

gas phase) are related to the distance between molecules. This observation is the same for all liquids. Experiments have shown that the surface properties of liquids change when solutes are added. This arises from the fact that the concentration of the solute may not be the same in the surface layer, as compared with the bulk concentration.

It is the cohesive forces that maintain water, for example, in liquid state at room temperature and pressure. This becomes obvious when one compares two different molecules, such as H_2O and H_2S. At room temperature and pressure, H_2O is a liquid while H_2S is a gas. This means that H_2O molecules interact with different forces more strongly (i.e., hydrogen bonds) to form a liquid phase. On the other hand, H_2S molecules exhibit much lower interactions, and thus are in a gas phase at room temperature and pressure.

2.2.1   SURFACE ENERGY

In any system (solid, liquid, solid–liquid, or $liquid_1$–$liquid_2$), if the surface area changes, then some molecules from the interior phase have to move to the surface. The state of *surface energy*, related in the latter case, has been described by the following classic example (Trevena, 1975; Adamson and Gast, 1997; Chattoraj and Birdi, 1984; de Gennes et al., 2003; Birdi, 1989, 1997, 2003a, 2003b). Consider the area of a liquid film that is stretched in a wire frame by an increment $d\mathbf{A}$, whereby the surface energy changes by ($\gamma\, d\mathbf{A}$) (Figure 2.4). Using these assumptions, one finds

$$\text{Surface tension} = \gamma \qquad (2.3)$$

$$\text{Change in area of the film} = d\mathbf{A} = l\,dx \qquad (2.4)$$

$$\text{Change in } x-\text{direction} = dx \qquad (2.5)$$

$$\mathbf{f}dx = \gamma d\mathbf{A} \qquad (2.6)$$

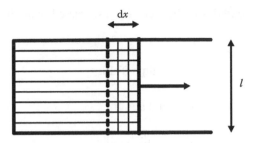

FIGURE 2.4   Surface film of a liquid.

or

$$\gamma = f\left(\frac{dx}{dA}\right)$$

$$= \frac{f}{2l} \tag{2.7}$$

where:
- **f**   is the opposing force
- dx is the change in displacement
- l is the length of the thin film

The quantity γ represents the force per unit length of the surface (mN/m=dyne/cm), and this force is defined as *surface tension* or *interfacial tension* (IFT). Experiments show that the molecules near or at the surface of a liquid are further apart than those in its interior. Surface tension, γ, is the differential change of free energy with change of surface area at constant temperature, pressure, and composition.

One may consider another example to describe the surface energy. Let us imagine that a liquid fills a container of the shape of a funnel. In the funnel, if one moves the liquid upward, then there will be an increase in surface area. This requires that some molecules from the bulk phase have to move into the surface area and create extra surface (=A_S). The magnitude of work required to do so will be (force×area)= (γ A_S). This is reversible work at constant temperature and pressure, from which one gets the increase in free energy of the system:

$$dG = \gamma A_S \tag{2.8}$$

Thus, the tension per unit length in a single surface, or surface tension, γ, is numerically equal to the surface energy per unit area. Then, G_S, the surface free energy per unit area is

$$G_S = \gamma = \left(\frac{dG}{dA_S}\right) \tag{2.9}$$

Under reversible conditions, the heat (**q**) associated with it gives the surface entropy, S_S:

$$dq = T \, dS_S \tag{2.10}$$

Combining these equations, we find that

$$\frac{d\gamma}{dT} = -Ss \tag{2.11}$$

Further, one finds

$$H_S = G_S + T_S \tag{2.12}$$

and one can also write for surface energy, E_S

$$E_S = G_S + T_S \, S_S \tag{2.13}$$

These relations give

$$E_S = \gamma - T\left(\frac{d\gamma}{dT}\right) \tag{2.14}$$

The quantity E_S has been found to provide more useful information on surface phenomena than any of the other thermodynamic quantities (Chattoraj and Birdi, 1984; Adamson and Gast, 1997). The sign of E_S is found to be always positive, since the quantities on the right-hand side in Equation 2.14 are always positive. Thus, S_S is the surface entropy per square centimeter of surface. This shows that to change the surface area of a liquid (or solid, as described in Chapter 2), there exists a surface energy (γ: surface tension), which one needs to consider.

The quantity γ means that to create 1 m² ($=10^{20}$ Å²) of new surface of water, one will need to use 72 mJ of energy. To transfer a molecule of water from the bulk phase (where it is surrounded by about 10 near neighbors by about 7 k_BT) ($k_BT = 4.12 \times 10^{-21}$ J) to the surface, one needs to break abot half of these hydrogen bonds (i.e., 7/2 $k_BT = 3.5$ k_BT). The free energy of transfer of one molecule of water (with area of 12 Å²) will thus be about 10^{-20} J (or about 3 k_BT). This magnitude is reasonable under these assumptions.

Further, it is found that somewhat similar consideration (with some modifications) is needed if one increases the surface area of a solid (e.g., by crushing [i.e., input of mechanical energy] or a similar case [such as fracture formation]). In the latter case, one needs to measure and analyze the surface tension of the solid (Chapter 4). Experiments have shown that the energy needed to crush a solid is related to the surface forces (i.e., solid surface tension). It thus becomes obvious that in many real-world situations (such as gas shale reservoirs), γ of both liquids and solids is needed to describe the surface chemistry of the system.

2.3 LAPLACE EQUATION FOR LIQUIDS (LIQUID SURFACE CURVATURE AND PRESSURE)

Experiments have shown that the most important parameter as regards the flow of a liquid in a porous medium is the capillary pressure. This has been extensively investigated in the literature for over a century (Scheludko, 1966; Goodrich et al., 1981; Birdi, 1999; Somasundaran, 2015). It is of interest to analyze a system in which a liquid comes into contact with a solid surface. Let us consider aspects in the field of wettability. Surely, everybody has noticed that water tends to rise near the walls of a glass container. This happens because the molecules of this liquid have a strong tendency to adhere to the glass. Liquids that wet the walls make concave surfaces (e.g., water/glass); those that do not wet them make convex surfaces (e.g., mercury/glass). Inside tubes with an internal diameter smaller than 2 mm, called *capillary tubes*, a wettable liquid forms a concave meniscus in its upper surface and tends to go up along the tube. In contrast, a nonwettable liquid forms a convex meniscus, and its level tends to go down. The amount of liquid attracted by the capillary rises till the forces that attract it balance the weight of the fluid column. The rising or lowering of the level of the liquids into thin tubes is named *capillarity* (*capillary force*). One notices that a liquid inside a large beaker is almost flat at the surface. However, the same liquid inside a fine tube will be found to be curved (Figure 2.5). The rise in height is found to be dependent on the radius of curvature. The capillary rise is higher in the narrow tube. This behavior is very important in everyday life. For example, in the case of oil or gas recovery, the most important characteristic is the pore size of the reservoir rock (which determines the capillary force). The physical nature of this phenomenon will be the subject of this section.

The mechanical equilibrium at liquid surfaces has been investigated for over a century. The liquid surface has been considered as a hypothetical stretched membrane, this membrane being termed the *surface tension* (Adamson and Gast, 1997; Chattoraj and Birdi, 1984; Birdi, 1989, 2003a, 2003b, 2016; Shou et al., 2014). In a real system undergoing an infinitesimal process, it can be written:

$$dW = p\,dV + p'\,dV' - \gamma\,dA \qquad (2.15)$$

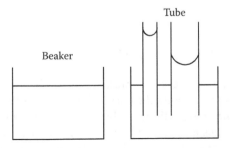

FIGURE 2.5 Surface of water inside a large beaker and in a narrow tube.

where:

dW   is the work done by the systems when a change in volume dV and dV', occurs

p and p' are pressures in the two phases α and β, respectively, at equilibrium

dA   is the change in interfacial area

The sign of the interfacial work is designated negative by convention (Chattoraj and Birdi, 1984; Adamson and Gast, 1997; Somasundaran, 2015). The fundamental property of liquid surfaces is that they tend to contract to the smallest possible area. This property is observed in the spherical form of small drops of liquid, in the tension exerted by soap films as they tend to become less extended, and in many other properties of liquid surfaces. In the case of oil or gas reservoirs, the recovery is primarily dependent on the interfacial forces. The pressure required to initiate flow of a liquid in porous media is related to the Laplace capillary pressure (Birdi, 2016). In the absence of gravity effects, these curved surfaces are described by the Laplace equation, which relates the mechanical forces as (Adamson and Gast, 1997; Chattoraj and Birdi, 1984; Birdi, 1997)

$$p - p' = \gamma \left(\frac{1}{r_1} + \frac{1}{r_2} \right) \tag{2.16}$$

$$= 2 \left(\frac{\gamma}{r} \right) \tag{2.17}$$

where:

r_1 and r_2 are the radii of curvature (in the case of an ellipse)

r is the radius of curvature for a spherical-shaped interface

It is a geometric fact that the surfaces for which Equation 2.11 holds are surfaces of minimum area. These equations thus give

$$dW = p d (V + V') - \gamma dA \tag{2.18}$$

$$= p dV^t - \gamma dA \tag{2.19}$$

where:

$p = p'$ for plane surface

V^t is the total volume of the system

It will be shown here that due to the presence of surface tension in liquids, there exists a pressure difference across the curved interfaces of liquids (such as drops or bubbles). This *capillary force* will be analyzed later. If one dips a tubing into water (or any fluid) and applies a suitable pressure, then a bubble is formed (Figure 2.4). This means that the pressure inside the bubble is greater than the atmosphere pressure. It thus becomes apparent that curved liquid surfaces induce effects that need special physico-chemical analyses in comparison with flat liquid surfaces. It must be noticed that in this system a *mechanical force* has induced a change on the surface of a liquid. This phenomenon is also called *capillary force*. Then, one may ask, does this also require similar consideration in the case of solids? Experiments show that solid surfaces also exhibit surface tension (Chapter 4). For example, to remove liquid that is inside a porous medium such as a sponge, one would need force equivalent to these capillary forces.

As seen in Figure 2.6, applying a suitable pressure, γP, to obtain a bubble of radius R, where the surface tension of the liquid is γ, produces a bubble.

Let us consider a situation in which one expands the bubble by applying pressure, P_{inside}. The surface area of the bubble will increase by dA, and the volume will increase by dV. This means that there are two opposing actions: expansion of volume and of surface area. The work done can be expressed in terms of that done against the forces of surface tension and that done in increasing the volume. At equilibrium, there will exist the following condition between these two kinds of work:

$$\gamma dA = \left(P_g - P_{liquid}\right)dV \tag{2.20}$$

where:

$$dA = 8\pi RdR\left(A = 4\pi R^2\right)$$
$$dV = 4\pi R^2 dR\left(V = 4/3\,\pi R^3\right)$$

Combining these relations gives

$$\gamma 8\pi R\ dR = \Delta P 4\pi R^2 dR \tag{2.21}$$

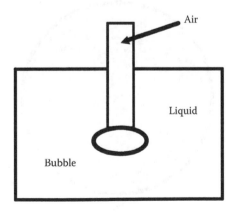

FIGURE 2.6   Formation of an air bubble in a liquid.

and

$$\Delta P = \frac{2\gamma}{R} \tag{2.22}$$

where $\Delta \mathbf{P} = (P_g - P_{liquid})$. Since the free energy of the system at equilibrium is constant ($\Delta G = 0$), then these two changes in the system are equal. If the same consideration is applied to the soap bubble, then the expression for $\Delta \mathbf{P}_{bubble}$ will be

$$\Delta \mathbf{P}_{bubble} = p_{inside} - p_{outside}$$

$$= \frac{4\gamma}{R} \tag{2.23}$$

since now there exist two surfaces and the factor of 2 is needed to consider this state. The pressure applied gives rise to work on the system, and the creation of the bubble gives rise to the creation of a surface area increase in the fluid. The Laplace equation relates the pressure difference across any curved fluid surface to the curvature, 1/radius, and its surface tension, γ. In those cases where nonspherical curvatures are present, one obtains the more universal equation

$$\Delta \mathbf{P} = \gamma \left(\frac{1}{R_1} + \frac{1}{R_2} \right) \tag{2.24}$$

It is also seen that in the case of spherical bubbles, since $R_1 = R_2$, this equation becomes identical to Equation 2.18. It is thus seen that in the case of a liquid drop in air (or gas phase), the Laplace pressure would be the difference between the pressure inside the drop, p_L, and the gas pressure, p_G (Figure 2.7):

$$p_L - p_G = \Delta P \tag{2.25}$$

$$= 2\gamma/radius \tag{2.26}$$

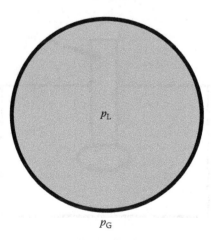

FIGURE 2.7   Liquid drop and ΔP (at a curved liquid–gas interface).

For example, in the case of a water drop with radius 2 μm, there will be ΔP of magnitude:

$$\Delta P = 2(72\,\text{mN/m})/2 \times 10^{-6}\,\text{m} = 72{,}000\,\text{N/m}^2 = 0.72\,\text{atm} \qquad (2.27)$$

The most important consequence of this is when there are two liquid drops in the vicinity. These will not be stable, due to the difference in the pressure. It is known that pressure changes the vapor pressure of a liquid. Thus, ΔP will affect the vapor pressure and lead to many consequences in different systems. This means that when two drops of different radii are in the vicinity, the smaller-sized drop will penetrate the larger drop (or bubble). Further, since the vapor pressure is dependent on the radius, this will give rise to asymmetry in different systems in which pores (of different size) are involved. The Laplace equation is useful for analyses in a variety of systems:

1. Bubbles or drops (raindrops, combustion engines, sprays, fog)
2. Oil/gas reservoirs and recovery processes
3. Groundwater movement in rocks
4. Biological phenomena (lung vesicles, blood cells (i.e., flow of blood through arteries), bacteria and viruses)

The important role of the Laplace equation and capillary pressure in various applications has been reported in the literature. For example, in oil/gas reservoirs, the pores are known to be very small (Appendix I). This gives rise to a major contribution to the pressure required to push fluids through the rocks due to the large magnitude of ΔP. Another important conclusion one may derive is that ΔP is larger inside a small bubble than in a larger bubble with the same γ. This means that when two bubbles meet, the smaller bubble will enter the larger bubble to create a new bubble (Figure 2.8). This phenomenon will have many important consequences in various systems (such as emulsion stability, lung alveoli, oil recovery, and bubble characteristics [such as in champagne and beer]). The same will be observed when two liquid drops contact each other; the smaller will merge into the larger drop.

Another important example is when two different-sized bubbles are present (Figure 2.7). In this system, initially, one has two bubbles of different curvature. After the tap is opened, (Figure 2.9) one finds that the smaller bubble shrinks, while

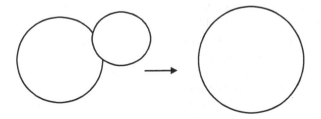

FIGURE 2.8 Coalescence of two bubbles with different radii in a liquid.

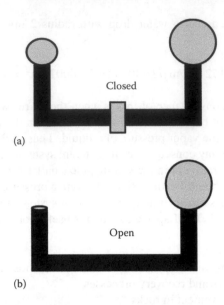

(a)

(b)

FIGURE 2.9 Equilibrium state of two bubbles of different radii: (a) valve closed; (b) valve opened.

the larger bubble (with lower ΔP) increases in size till equilibrium is reached (when the curvature of the two bubbles becomes equal in magnitude).

It has been observed that a system with bubbles of varying sizes collapses faster than when bubbles (or liquid drops) are of exactly the same size. Another major consequence is observed in oil-recovery phenomena (Chapter 5). Oil production takes place (in general) by applying pressure using gas or water (or other) injection. When gas or water injection is applied, then where there are small pores, the pressure needed will be higher than in the large pore zone. Thus, the gas or water will bypass the small pore zone and leave the oil behind (at present more than 30%–50% of oil in place is not recovered under normal production methods). This is obviously a great challenge for surface and colloid chemists in the future. Enhanced oil recovery (EOR) processes are mainly related to these surface phenomena (Chapter 5). Similar characteristics will be observed in shale reservoirs. In the shale reservoirs, the basic surface chemical principles will be the same.

In industry, one can monitor the magnitude of surface tension of a liquid with the help of bubble pressure. Air bubbles are pumped through a capillary into the solution. The pressure measured is calibrated to known surface tension solutions; thus, by using a suitable computer, one can estimate surface tension values very accurately. Commercial apparatus is available to monitor surface tension.

The consequence of Laplace pressure is given as an example. One important example is that when a small drop comes into contact with a larger drop, the former will merge into the latter. Another aspect is that vapor pressure over a curved liquid surface, p_{cur}, will be larger than on a flat surface, p_{flat}. A relation between pressure over curved and flat liquid surfaces was derived (Kelvin equation) (Adamson and Gast, 1997; Birdi, 2016):

$$\ln\left(p_{cur}/p_{flat}\right) = \left(v_L/R\ T\right)\left(2\gamma/R_{cur}\right) \qquad (2.28)$$

where:

p_{cur} and p_{flat} are the vapor pressures over curved and flat surfaces, respectively

R_{cur} is the radius of curvature

v_L is the molar volume

This relationship thus suggests that if liquid is present in a porous material (such as cement or oil/gas reservoirs), then a difference in vapor pressure exists between two pores of different radii. In cement, this gives rise to drastic differences in the solidification process (arising from the asymmetric vapor pressure in different-size pores). This problem is reduced by adding SAS, which reduces the magnitude of ΔP. A similar consequence of vapor pressure exists when two solid crystals of different sizes are concerned. The smaller-sized crystal will exhibit higher vapor pressure, and will also result in a faster solubility rate. Further, the transition from water vapor in clouds to raindrops is not as straightforward as it might seem. The formation of a large liquid raindrop requires that a certain number of water molecules in the clouds have formed nuclei. The nuclei or embryo will grow larger to form a drop.

2.4 CAPILLARY RISE (OR FALL) OF LIQUIDS

The behavior of liquids in narrow tubes is one of the most common examples in which capillary forces are involved. Experiments have shown that this phenomenon plays an important role in many different parts of everyday life and technology (Birdi, 2014; Somasundaran, 2015). In fact, liquid curvature is one of the most important physical surface properties and requires attention in most application areas of this science. These applications range from the blood flow in the veins to oil recovery in the reservoir. For example, fabric properties are also governed by capillary forces (that is, wetting, etc.). The sponge absorbs water or other fluids when the capillary forces push the fluid into the pores. This is also called the *wicking* process (as in candlewicks). In fact, one may compare the flow of liquid in a sponge to that in an oil/gas reservoir.

As an example, let us analyze a system in which a narrow capillary circular tube is dipped into a liquid. The liquid is found to rise in the capillary when the fluid *wets* the capillary (such as water and glass or water and metal). The *curvature* of the liquid inside the capillary will lead to a pressure difference between this state and the relatively flat surface outside the capillary (Figure 2.10).

The fluid with surface tension γ wets the inside of the tube, which gives an equilibrium of capillary forces. However, if the fluid is *nonwetting* (such as mercury in glass), then one finds that the fluid *falls*. This arises from the fact that mercury does not wet the tube. Capillary forces arise from the difference between the attraction of the liquid molecules to each other and the attraction of the liquid molecules to those

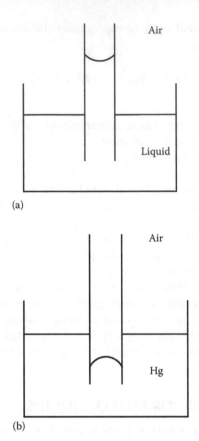

(a)

(b)

FIGURE 2.10   (a) Rise of water in a glass capillary; (b) fall of mercury.

of the capillary tube. The fluid rises inside the narrow tube to a height, **h**, till the surface tension forces balance the weight of the fluid. This equilibrium gives the relation

$$\gamma 2\pi \mathbf{R} = \text{surface tension force} \tag{2.29}$$

$$= \rho_L g_g h\pi \mathbf{R}^2 = \text{fluid weight inside the capillary} \tag{2.30}$$

where:
  γ is the surface tension of the liquid
  R   is the radius of curvature

In the case of narrow capillary tubes, less than 0.5 mm, the curvature can be safely set equal to the radius of the capillary tube. The fluid will rise inside the tube to compensate for the surface tension force; thus, at equilibrium, we get

$$\gamma = 2\mathbf{R}\rho_L \mathbf{g}\mathbf{h} \tag{2.31}$$

where:

 ρ_L is the density of the fluid
 g   is the acceleration of gravity
 h   is the rise in the tube

Water ($\gamma = 72.8$ mN/m; $\rho = 1000$ kg/m³; **g** $=9.8$ m/s²) will rise to a height of

- 0.015 mm in a capillary of radius 1 m
- 1.5 mm in a capillary of radius 1 cm
- 14 cm in a capillary of radius 0.1 mm

This data shows that the capillary pressures will be significant in porous materials.

This ideal relation needs corrections for higher accuracy. Tables for such corrections are found in the published literature (Adamson and Gast, 1997). This phenomenon is important for plants with very tall stems for which water is needed for growth.

However, if in some particular case the contact angle, θ, is not zero, the equation will become

$$\gamma = 2R\rho_L gh\left(\frac{1}{\cos(\theta)}\right)$$
(2.32)

It is seen that when the liquid wets the capillary wall, the magnitude of the contact angle is equal to 0, and $\cos(0) = 1$. In the case of mercury, the contact angle is 180°, since it is a nonwetting fluid (see Figure 2.8). Since $\cos(180) = -1$, the sign of **h** in Equation 2.32 will be negative. This means that mercury will show a drop in height in glass tubing. Hence, the rise or fall of a liquid in tubing will be governed by the sign of $\cos(\theta)$.

Thus, capillary forces will play an important factor in all systems in which liquids are present in a porous environment.

A similar result can also be derived by using the Laplace Equation 2.21 (1/radius $= 1/R$):

$$\Delta P = 2\gamma/R$$
(2.33)

The liquid rises to a height **h**, the system achieves equilibrium, and the following relation is found.

$$2\gamma/R = hg_g\rho_L$$
(2.34)

This can be rewritten as

$$\gamma = 2R\rho_L g_g h$$
(2.35)

It is thus seen that the various surface forces are responsible for the capillary rise. The lower the surface tension, the lower the height of the column in the capillary.

The magnitude of γ is determined from the measured value of **h** for a fluid with known ρ_L. The magnitude of **h** can be measured directly by using a suitable device (e.g., a photographic image).

Further, it is known that in the real world, capillaries or pores are not always circular in shape. In fact, it is thought that in oil reservoirs, the pores are more triangular or square than circular. In this case, one can measure the rise in capillaries of other shapes, such as rectangular or triangular (Birdi et al., 1988; Birdi, 1997, 2003a, 2003b).

These studies provide much important information for oil-recovery or water-treatment systems. In any system in which fluid flows through a porous material, one would expect that capillary forces would be one of the most dominant factors. Further, it is known that the vegetable world is dependent on capillary pressure (and osmotic pressure) to bring water up to the higher parts of plants.

2.5 BUBBLE (OR FOAM) FORMATION

It is interesting to notice that one finds some systems that are useful in a great variety of aspects. The structure and formation of bubbles from soap solutions are one of these special systems. One phenomenon that everyone can quickly recognize is soap bubbles, which one has observed since childhood. The formation of foam bubbles, from beers or washing machines to the coasts of lakes and oceans, is another common experience. The formation and stability of bubbles and foams have been extensively investigated in the literature (Lovett, 1994; Scheludko, 1966; Rubinstein and Bankoff, 2001; Birdi, 1997, 2016). It is also known that soap bubbles are extremely thin and unstable. Despite this, under special conditions, it is possible to keep soap bubbles for a long time, which allows one to study their physical properties (thickness, composition, conductivity, spectral reflection, etc.).

Foam or bubble formation:

- Shaking pure water: no foam or bubbles are formed
- Shaking a soap solution: foam and bubbles are formed

The study of bubbles is essential, since this allows one to understand the structure of molecules at liquid surfaces (Chapter 7). The thickness of a bubble is in most cases over hundreds of micrometers in the initial state. The film consists of a bilayer of detergent, which contains the solution. The film thickness decreases with time due to

- Drainage of fluid away from the film
- Evaporation

Therefore, the stability and lifetime of such thin films will be dependent on these different characteristics. This is found from the fact that as an air bubble is blown under the surface of a soap or detergent solution, it will rise up to the surface. It may remain at the surface, if the speed is slow, or it may escape into the air as a soap bubble. Experiments show that a soap bubble consists of a very thin liquid film with an iridescent surface. But as the fluid drains away and the thickness decreases, the latter approaches the equivalent of barely two surfactant molecules plus a few molecules of water.

It is worth noting that the limiting thickness is of the order of two or more surfactant molecules. This means that one can see with the naked eye molecular-size structures of thin liquid films (if curved).

As the air bubble enters the surface region, the soap molecules are pushed up, and as the bubble is detached, it leaves as a thin liquid film with the following characteristics (as found from various measurements):

- A bilayer of soap (approximately 200 Å thick) on the outer region contains the aqueous phase.
- The thickness of the initial soap bubble is some micrometers.
- The thickness decreases with time, and one starts to observe rainbow colors, as the reflected light is of the same wavelength as the thickness of the bubble (a few hundred Ångstroms).
- The thinnest liquid films consist mainly of a bilayer of surface-active substance (such as soap = 50 Å) and some layers of water. Light interference and reflection studies show many aspects of these thin liquid films.

The iridescent colors of the soap bubble arise from the interference of reflected light waves. The reflected light from the outer surface and the inner layer give rise to this interference effect. The rainbow colors are observed as the bubble thickness decreases due to the evaporation of water. Thicker films reflect more red light, and therefore, one observes blue-green colors. Thinner films cancel out the yellow wavelengths, and a blue color is observed. As the thickness approaches the wavelength of light, all colors are cancelled out, and a so-called black (or gray) film is observed. This corresponds to 25 nm (250 Å) (Scheludko, 1966; Birdi, 2016). The transmitted light, I_{tr}, is related to the incident, I_{in}, and the reflected intensity, I_{re}:

$$I_{tr} = I_{in} - I_{tr} \tag{2.36}$$

Daylight consists of different wavelengths of colors (Blomberg, 2007):

Red	680 nm
Orange	590 nm
Yellow	580 nm
Green	530 nm
Blue	470 nm
Violet	405 nm

Slightly thicker soap films (ca. 1500 nm) sometimes look golden. In the thinning process, different colors are cut off. Thus, if the blue color is cut off, the film looks amber to magenta.

Some recipes for soap bubble solutions follow.

In general, one may use a detergent solution (1–10 g/L detergent concentration). However, to produce stable bubbles (which means slow evaporation rates, stability to vibration effects, etc.), some additives have been used. A common recipe is

- Detergent (dishwashing): 1–10 g/L
- Glycerin: less than 10%
- Water: rest

Another recipe has the following composition (Lovett, 1994):

$$100 \text{ g glycerine} + 1.4 \text{ g triethanolamine} + 2 \text{ g oleic acid}$$

Bubble film stability can be described as follows:

- Bubble film: evaporation of water
- Flow of water away from the film
- Stability of the bubble film

Containing the bubble in a closed container can reduce evaporation. One also finds that in such a closed system, the bubbles remain stable for a very long time. The drainage of water away from the film is dependent on the viscosity of the fluid. Therefore, additives such as glycerin (or other thickening agents [polymers]) assist in the stability.

2.6 MEASUREMENT OF SURFACE TENSION OF LIQUIDS

In the case of liquids, the surface forces (surface tension) are a basic physical property. It is thus essential that the surface tension data is known, to describe the surface properties. It is apparent that the measurement of the surface tension of liquids is an important analysis. The method that one may choose depends on the system and the experimental conditions (as well as the accuracy needed). For example, if the liquid is water (at room temperature), then the method will be different than if the system is molten metal (at very high temperature, ca. 500°C). These different systems and methods will be explained in this section.

2.6.1 LIQUID DROP WEIGHT AND SHAPE METHOD

The formation of liquid drops when flow occurs through thin tubes is a common daily phenomenon. The drop formed when liquid flows through a circular tube is shown in Figure 2.11.

In many processes (such as oil recovery, blood flow, and underground water), one encounters liquid flow through thin (micrometer diameter) noncircular tubes or pores. Studies in the literature address these systems.

In other contexts, for example, liquid drop formation in an ink-jet nozzle, the technique falls into the class of scientifically challenging technology. An ink-jet printer

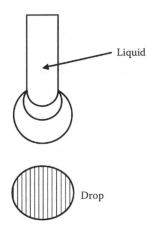

FIGURE 2.11   Drop formation of a liquid at the end of a tube.

demands such quality that this branch of drop-on-demand technology is the subject of much industrial research and technology. All combustion engines are controlled by oil-drop formation, flow, and evaporation characteristics. The important role of capillary forces is obvious in such systems. As the liquid drop grows larger, it will at some stage break off from the tube (due to the force of gravity being larger than the surface force). This will correspond to the maximum weight of the drop that can remain hanging.

The equilibrium state, in which the weight of the drop is exactly equal to the detachment surface energy, is given as

$$m_m \, \mathbf{g} = 2\pi \mathbf{R}\gamma \tag{2.37}$$

where:
   m_m is the weight of the detached drop
   $\mathbf{R}$ is the radius of the tubing

A simple method is to *count* a number of drops (e.g., 10 or more) and measure their weight.

One may also use a more convenient method whereby a fluid is pumped and the drops are collected and weighed. Since in some systems (solutions) there may be kinetic effects, one must be careful to keep the flow as slow as possible. This method is very useful in studying systems that one finds in the phenomena of daily life, such as oil flow and the flow of blood cells through arteries. In cases where the volume of fluid available is limited, one may use this method with advantage. By decreasing the diameter of the tubing, one can work with quantities of fluid lower than 1 mL. This may be useful in the case of such systems as eye fluids.

One can determine the magnitude of γ from either the maximum weight or the shape of the drop.

2.6.1.1   Maximum Weight Method

The "detachment" method is based on detaching a body from the surface of a liquid that wets the body. It is necessary to overcome the same surface tension forces that operate when a drop is broken away. The liquid attached to the solid surface on detachment creates the following surfaces:

- Initial stage: liquid attached to solid
- Final stage: liquid separated from solid

In the process from the initial to the final stage, the liquid molecules that were near the solid surface have been moved away and are now near other liquid molecules. This requires energy, and the force required to make this happen is proportional to the surface area of contact and to the surface tension of the liquid. However, the advantage of this method is that it makes it possible to choose the most convenient form and size of the body (platinum rod, ring, or plate) so as to enable the measurement to be carried out rapidly, but without any detriment to its accuracy. The "detachment" method has found application in the case of liquids whose surface tension changes with time.

2.6.1.2   Shape of the Liquid Drop (Pendant Drop Method)

This method is one of the most versatile for a range of phenomena.

The liquid drop forms as it flows through a tube (Figure 2.12). At a stage just before it breaks off, the shape of the *pendant drop* has been used to estimate γ. The drop shape is photographed, and from the diameters of the shape, one can accurately determine γ.

The parameters needed are as follows. A quantity pertaining to the ratio of two significant diameters:

$$S = \frac{d_s}{d_e} \tag{2.38}$$

Profile of a pendant drop

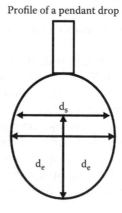

FIGURE 2.12   Pendant drop of liquid (shape analysis).

where:

    d_e is the equatorial diameter

    d_s is the diameter at a distance d_e from the tip of the drop (Figure 2.11)

The relation between γ, d_e, and **S** is

$$\gamma = \frac{\rho_L g d_e^2}{H} \tag{2.39}$$

where:

    ρ_L is the density of the liquid.

    H   is related to **S**, but the values of $1/H$ for varying **S** have been obtained from experimental data.

For example, when **S**$=0.3$, $1/H=7.09837$, while when **S**$=0.6$, $1/H=1.20399$. Accurate mathematical functions have been used to estimate $1/H$ for a given d_e value (Adamson and Gast, 1997; Frohn and Roth, 2000; Birdi, 2003a, 2003b). The accuracy (0.1%) is satisfactory for most systems, especially when experiments are carried out under extreme conditions (such as high temperatures and pressures).

The pendant drop method is very useful under specific conditions:

1. Technically, only a drop (a few microliters) is required. For example, eye fluid can be studied.
2. It can be used under very extreme conditions (very high temperature or corrosive fluids).
3. Under very high pressure and temperatures.

Oil reservoirs are found typically at 100°C and 300 atm pressure. The surface tension of such systems can be conveniently studied by using high-pressure and high-temperature cells with optical clear windows (sapphire windows 1 cm thick, up to 2000 atm). For example, γ of inorganic salts at high temperatures (ca. 1000°C) can be measured using this method. The variation of surface tension can be studied as a function of various parameters (temperature and pressure, additives such as gas, etc.).

2.6.2 PLATE METHOD (WILHELMY)

The methods so far discussed have required more or less tabular solutions, or else correction factors to the respective "ideal" equations. Further, if one needs to make continuous measurements, it is not easy to use some of these methods (such as the capillary rise or bubble method). The most useful method of measuring the surface tension is the well-known *Wilhelmy* plate method. If a plate-shaped piece of metal is dipped into a liquid, the surface tension forces will be found to give rise to a tangential force (Figure 2.13). This is because a new contact phase is created between the plate and the liquid.

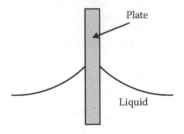

FIGURE 2.13   Wilhelmy plate in a liquid (plate dimensions: length = Lp, width = Wp).

The total weight measured, W_{total}, is

$$W_{total} = \text{weight of the plate} + \gamma\,(\text{perimeter}) - (\text{updrift}) \tag{2.40}$$

The surface force will act along the perimeter of the plate (i.e., length (L_p) + width (W_p)). The plate is often very thin (less than 0.1 mm) and made of platinum, but plates made of glass, quartz, mica, and filter paper can be used. The forces acting on the plate consist of the gravity and surface tension downward, and buoyancy due to displaced water upward. For a rectangular plate of dimensions L_p and W_p and of material density ρ_p, immersed to a depth h_p in a liquid of density ρ_L, the net downward force, F, is given by the following equation (i.e., weight of plate + surface force ($\gamma \times$ perimeter of the plate – upward drift):

$$\mathbf{F} = \rho_p \mathbf{g}\left(L_p W_p t_p\right) + 2\gamma\left(t_p + W_p\right)\left(\cos(\theta)\right)\rho_L \mathbf{g}\left(t_p W_p h_p\right) \tag{2.41}$$

where:
  γ is the liquid surface tension
  θ is the contact angle of the liquid on the solid plate
  $\mathbf{g}$ is the gravitational constant

The weight of the plate is constant and can be tared. If the plate used is very thin (i.e., $t_p \ll W_p$) and the updrift is negligible (i.e., h_p is almost zero), then γ is found from

$$\gamma = \frac{(\mathbf{F})}{(2 W_p)} \tag{2.42}$$

The sensitivity of γ using these procedures has been found to be very high (± 0.001 dyne/cm (mN/m)) (Birdi, 2003a). The change in surface tension (surface pressure = Π) is then determined by measuring the change in $\mathbf{F}$ for a stationary plate between a clean surface and the same surface with a monolayer present. If the plate is completely wetted by the liquid (i.e., $\cos(\theta) = \cos(0) = 1$), the surface pressure is obtained from the equation

$$\Pi = \left[\frac{\Delta F}{2} \left(t_p + W_p \right) \right]$$

$$= \frac{\Delta F}{2} W_p, \text{ if } W_p \gg t_p \qquad (2.43)$$

Thus, by using very thin plates with thickness 0.1–0.02 mm, one can measure surface tension with very high sensitivity. In practice, by using very thin platinum plates of well-known dimensions (length = 1.00 or 2.00 cm), one can calibrate the apparatus with pure liquids, such as water and ethanol. The buoyancy correction is made negligible by using a very thin plate and dipping the plate as little as possible. The wetting of water on platinum plate is achieved very satisfactorily by using commercially available plates, which are roughened. The latter property gives rise to almost complete wetting, that is, $\theta = 0$. The force is in this way determined by measuring the changes in the mass of the plate, which is directly coupled to a sensitive electro-balance.

2.7   SURFACE TENSION DATA OF SOME TYPICAL LIQUIDS

It is useful to analyze some data of surface tension of liquids. This would allow one to understand the relation between the structure of a molecule and its surface tension. Some typical values of surface tension of different liquids are given in Table 2.1. A brief analysis of this data is given in this section.

This data needs some comments to explain the differences in surface tension data and molecular structure (Adamson and Gast, 1997; Chattoraj and Birdi, 1984; Birdi, 1997, 2003a, 2003b, 2016). For example, the surface tension of mercury is high because it is a liquid metal with a very high boiling point (at room temperature and pressure). This observation indicates that much energy is needed to break the bonds between mercury atoms to cause evaporation, and hence, the value of surface tension is high. Similarly, γ of NaCl (mN/m) as a liquid (at high temperature and 1 atm pressure) is also very high, as expected. The same is found for metals in the liquid state (at high temperature). Shorter-chain alkanes exhibit lower values of surface tension than longer chains.

For alkanes, the magnitude of γ increases by 1.52 mN/m per two CH_2 groups when the alkyl chain length increases from 10 to 12 (n-decane = 23.83 mN/m; n-dodecane = 25.35 mN/m).

n-Hexane	18.43
n-Heptane	20.14
n-Octane	21.62
n-Hexadecane	27.47

TABLE 2.1
Surface Tension Values of Some Common Liquids

Liquid	Surface Tension (20°C; mN/m [dyne/cm])
1,2-Dichloroethane	33.3
1,2,3-Tribromopropane	45.4
1,3,5-Trimethylbenzene (mesitylene)	28.8
1,4-Dioxane	33.0
1,5-Pentanediol	43.3
1-Chlorobutane	23.1
1-Decanol	28.5
1-Nitropropane	29.4
1-Octanol	27.6
Acetone (2-propanone)	25.2
Aniline	43.4
2-Aminoethanol	48.9
Anthranilic acid ethyl ester	39.3
Anthranilic acid methyl ester	43.7
Benzene	28.9
Benzyl alcohol	39.0
Benzylbenzoate (BNBZ)	45.9
Bromobenzene	36.5
Bromoform	41.5
Butyronitrile	28.1
n-Propanol	23.39
n-Butanol	24.37
n-Pentanol	25.33
n-Hexanol	25.90
Carbon disulfide	32.3
Quinoline	43.1
Chlorobenzene	33.6
Chloroform	27.5
Cyclohexane	24.9
Cyclohexanol	34.4
Cyclopentanol	32.7
p-Cymene	28.1
Decalin	31.5
Dichloromethane	26.5
Diiodomethane	50.8
1,3-Diiodopropane	46.5
Diethylene glycol	44.8
Dipropylene glycol	33.9
Dipropylene glycol monomethyl ether	28.4
Dodecylbenzene	30.7
Ethanol	22.1
Ethylbenzene	29.2

TABLE 2.1 (CONTINUED)
Surface Tension Values of Some Common Liquids

Liquid	Surface Tension (20°C; mN/m [dyne/cm])
Ethyl bromide	24.2
Ethylene glycol	47.7
Formamide	58.2
Fumaric acid diethyl ester	31.4
Furfural (2-furaldehyde)	41.9
Glycerol	64.0
Ethylene glycol monoethyl ether (ethyl cellosolve)	28.6
Hexachlorobutadiene	36.0
Iodobenzene	39.7
Isoamylchloride	23.5
Isobutylchloride	21.9
Isopropanol	23.0
Isopropylbenzene	28.2
Isovaleronitrile	26.0
m-Nitrotoluene	41.4
Mercury	425.4
Methanol	22.7
Methyl ethyl ketone (MEK)	24.6
Methylnaphthalene	38.6
N,N-dimethyl acetamide (DMA)	36.7
N,N-dimethyl formamide (DMF)	37.1
N-methyl-2-pyrrolidone	40.7
n-Decane	23.8
n-Dodecane	25.3
n-Heptane	20.1
n-Hexadecane	27.4
n-Hexane	18.4
n-Octane	21.6
n-Nonane	22.4
n-Tetradecane	26.5
n-Undecane	24.6
n-Butylbenzene	29.2
n-Propylbenzene	28.9
Nitroethane	31.9
Nitrobenzene	43.9
Nitromethane	36.8
o-Nitrotoluene	41.5
Perfluoroheptane	12.8
Perfluorohexane	11.9
Perfluorooctane	14.0
Phenylisothiocyanate	41.5

(Continued)

TABLE 2.1 (CONTINUED)
Surface Tension Values of Some Common Liquids

Liquid	Surface Tension (20°C; mN/m [dyne/cm])
Phthalic acid diethyl ester	37.0
Polyethylene glycol 200 (PEG)	43.5
Polydimethylsiloxane	19.0
Propanol	23.7
Pyridine	38.0
3-Pyridylcarbinol	47.6
Pyrrole	36.6
sym-Tetrabromoethane	49.7
tert-Butylchloride	19.6
sym-Tetrachloromethane	26.9
Tetrahydrofuran (THF)	26.4
Thiodiglycol	54.0
Toluene	28.4
Tricresylphosphate (TCP)	40.9
Water	72.8
o-Xylene	30.1
m-Xylene	28.9
a-Bromonaphthalene	44.4
a-Chloronaphthalene	41.8
Mercury	425.4
Sodium (100°C)	206.0

For alcohols, the magnitude of γ changes by $23.7 - 22.1 = 1.6$ mN/m per CH_2 group. This is based on the γ data of ethanol (22.1 mN/m) and propanol (23.7 mN/m). These observations indicate the molecular correlation between bulk forces and surface forces (surface tension, γ) for homologous series of substances.

2.8 EFFECT OF TEMPERATURE AND PRESSURE ON SURFACE TENSION OF LIQUIDS

All natural processes are found to be dependent on the temperature and pressure effects on any system under consideration. For example, oil and gas reservoirs are generally found under high temperature (ca. 100°C) and pressure (over 200 atm). In the shale gas fracking process, the chemical process takes place under high pressure and temperature. In the fracking process, the liquid–solid–gas interfaces are under these conditions. Mankind is aware of great variations of both temperature (sun) and pressure (earthquakes) in the natural phenomena affecting the earth. Even on the surface of the earth, the temperature varies between −50 and +50°C. On the other hand, the mantle of the earth increases in temperature and pressure as one goes from its surface to the center of the earth (about 5000 km). The inner core of the earth is at about 6000°C and at very high pressure (the approximate rate of increase is ca. 100 atm per 1 km depth).

The surface tension is related to the internal forces in the liquid (surface), and one must thus expect it to show a relationship to the internal energy. Further, it is found that surface tension always decreases with increasing temperature. Surface tension, γ, is a quantity that can be measured accurately and applied in the analysis of all kinds of surface phenomena. If a new surface is created, then in the case of a liquid, molecules from the bulk phase must move to the surface. The work required to create extra surface area, dA_S, is given as

$$dG_S = \gamma dA_S \tag{2.44}$$

The surface free energy, G_S, per unit area is given as

$$G_S = \gamma = \left(\frac{dG}{dA_S} \right)_{T,P} \tag{2.45}$$

Hence, the other thermodynamic surface quantities will be
Surface entropy, S_s:

$$S_S = -\left(\frac{dG_S}{dT} \right)_P \tag{2.46}$$

$$= -\left(\frac{d\gamma}{dT} \right) \text{ (always a positive quantity)} \tag{2.47}$$

Surface enthalpy, H_S:

$$H_S = G_S + TS_S \tag{2.48}$$

All natural processes are dependent on the effects of temperature and pressure. For instance, oil reservoirs are found under high temperatures (ca. 80°C) and pressure (around 100–400 atm, depending on the depth). Thus, one must investigate such systems under these parameters. This is related to the fundamental equation for the free energy, G, and to the enthalpy, H, and entropy, S, of the system:

$$G = H - TS \tag{2.49}$$

This equation relates the essential thermodynamic quantities of any system. The molecular forces that stabilize liquids will be expected to decrease as the temperature increases. Experiments also show that in all cases, surface tension decreases with increasing temperature.

The surface entropy of liquids is given by $(-d\gamma/dT)$. This means that the entropy is *positive* at higher temperatures (the sign of $d\gamma/dT$ is always negative for all liquids).

The rate of decrease of surface tension with temperature is found to be different for different liquids (Appendix I), which supports this description of liquids.

For example, the surface tension of water, γ, is

- At 5°C, 75 mN/m
- At 25°C, 72 mN/m
- At 90°C, 60 mN/m

This shows that the magnitude of γ of any liquid decreases with temperature, and this property is found to be just as specific as other physical properties of matter, such as boiling point, melting point, and heat of vaporization critical point (pressure and temperature). It is found that surface tension is related to the last of these. Extensive accurate γ data for water was fitted to the following equation (Birdi, 2016):

$$\gamma = 75.69 - 0.1413t_C - 0.0002985t_C^2 \qquad (2.50)$$

where t_C is in degrees Centigrade. This equation gives the value of γ at 0°C as 75.69 mN/m. The value of γ at 50°C is found to be (75.69– $0.1413 \times 50 - 0.0002985 \times 50 \times 50) = 60$ mN/m. Further, these data show that γ of water decreases with temperature from 25°C to 60°C: $(72-60)/(90-25) = 0.19$ mN/m°C. The differences in the surface entropy give information on the structures of different liquids. It is also observed that the effect of temperature will be lower for liquids with higher boiling point (such as mercury) than for low-boiling liquids (such as n-hexane). Actually, there exists a correlation between surface tension and heat of vaporization (or boiling point). In fact, many systems even show big differences when winter and summer months are compared (such as raindrops, sea waves, and foaming in natural environments).

Different thermodynamic relations have been derived that can be used to estimate the surface tension at different temperatures. Especially, straight-chain alkanes have been extensively analyzed. The data shows that there exists a simple correlation between surface tension, temperature, and n_C. This allows one to estimate the value of γ at any temperature of a given alkane. This observation has many aspects in applied industry. It allows one to estimate the magnitude of surface tension of an alkane at the required temperature. Further, extensive quantitative analyses of the change of surface tension of liquids (alkanes) under given experimental conditions are available (Appendix III).

2.8.1 Heat of Liquid Surface Formation and Evaporation

All matter (solid or liquid) is stabilized by the forces between the molecules (at a given temperature and pressure). The magnitude of forces is related to the distance between the molecules. Liquid structure is stabilized by different molecular forces. In oil and gas reservoirs, one has systems at high temperature and pressure. This requires knowledge of the system in these conditions. To understand how liquids are stabilized, several attempts have been made to relate the surface tension of a liquid to the latent heat of evaporation. This has been attempted based on some simple geometrical molecular packing of liquids. A simple theory was proposed (Stefan, 1886; Chattoraj and Birdi, 1984) that when a molecule is brought to the surface of a liquid from the interior, the work done in overcoming the attractive force near the surface should be related to the work expended when it escapes into the scarce (less dense) vapor phase (Adamson and Gast, 1997; Birdi, 1997, 2002). It was suggested that the first quantity should be approximately half of the second. According to the Laplace theory of capillarity, the attractive force acts only over a small distance, equal to the radius of the sphere (see Figure 1.1), and in the interior, the molecule is attracted equally in all directions and experiences no resultant force. At the surface, it experiences a force due to the liquid in the hemisphere, and half the total molecular attraction is overcome in bringing it there from the interior. Accordingly, the energy necessary to bring a molecule from the bulk phase to the surface of a liquid should be half the energy necessary to bring it entirely into the gas phase (latter step = heat of evaporation). From these simple geometrical considerations, one knows that a sphere can be surrounded by 12 molecules of the same size. This corresponds to the most densely packed top (surface) monomolecular layer half filled, and the next layer completely filled, next to a very dilute gas phase (the distance between gas molecules is approximately 10 times greater than in liquids or solids). This indicates that intermolecular forces in liquids will be weaker than those in solids by a few orders of magnitude, as is also found experimentally.

The ratio of the enthalpy of surface formation to the enthalpy of vaporization, $h_s:h_{vap}$, for various substances is given in Table 2.2. This data shows that substances with nearly spherical molecules have ratios near 1/2, while substances with a polar group on one end give a much smaller ratio. This difference indicates that the latter

TABLE 2.2
Enthalpy of Surface Formation and Ratio to Enthalpies of Vaporization

Molecule (liquid)	h_s/h_{vap}
Hg	0.64
N_2	0.51
O_2	0.5
CCl_4	0.45
C_6H_6	0.44
Diethyl ether	0.42
ClC_6H_5	0.42
Methyl formate	0.40
Ethyl acetate	0.4
Acetic acid	0.34
H_2O	0.28
C_2H_5OH	0.19
CH_3OH	0.16

Note: h_s is given in erg/molecule.

molecules are oriented with the nonpolar end toward the gas phase and the polar end toward the liquid. In other words, molecules with dipoles would be expected to be oriented perpendicularly at the gas/liquid interfaces.

In fact, any deviation from Stefan's law is an indication that the surface molecules are oriented differently than in the bulk phase. This observation is useful to understand surface phenomena.

As an example, one may proceed with this theory and estimate the surface tension of a liquid with data on its heat of evaporation. The number of near neighbors of a surface molecule will be about half (6 = 12/2) those in the bulk phase (12 neighbors). It is now possible to estimate the ratio of the attractive energies in the bulk and at the surface, per molecule. For example, in the case of a liquid such as CCl_4,

$$\text{Molar energy of vaporization} = \Delta U_{vap} \qquad (2.51)$$

$$= \Delta h_{vap} - \textbf{RT}$$

$$= 34{,}000 \text{ J/mol} - 8.315 \text{ J/K/mol} \left(298 \text{ K}\right)$$

$$= 31{,}522 \text{ J/mol} \qquad (2.52)$$

$$\text{Energy change per molecule} = \frac{31{,}522 \text{ J/mol}}{6.023 \times 10^{23} /\text{mol}} = 5.23 \times 10^{-20} \text{ J} \qquad (2.53)$$

If it is assumed that about half the energy is gained when a molecule is transferred to the surface, one obtains

$$\text{Energy per molecule at surface} = 5.23 \times 10^{-20} \left(1/2\right) \text{J}$$

$$= 2.6 \times 10^{-20} \text{ J} \qquad (2.54)$$

The molecules at the surface occupy a certain value of area, which can be estimated only roughly, as follows.

$$\text{Density of } CCl_4 = 1.59 \text{ g/cm}$$

$$\text{Molar mass} = 12 + 4\left(35.5\right) = 154 \text{ g/mol}$$

$$\text{Volume per mol} = \frac{154}{1.59} = 97 \text{ cm}^3/\text{mol}$$

$$\text{Volume per molecule} = \frac{97 \times 10^{-6} \text{ m}^3/\text{mol}}{6.023 \times 10^{23} /\text{mol}}$$

$$= 1.6 \times 10^{-28} \text{ m}^3$$

The radius of a sphere (volume $= 4/3 \; \Pi \; R^3$) with this magnitude of volume $= (1.6 \times 10^{-28}/(4/3 \; \Pi))^{1/3}$

$$= 3.5 \times 10^{-10} \, m$$

$$\text{Area per molecule} = \Pi R^2$$

$$= \Pi \left(3.5 \times 10^{-10}\right)^2 = 38 \times 10^{-20} \, m^2$$

$$\text{Surface tension (calculated) for } CCl_4 = \frac{2.6 \times 10^{-20} \, J}{38 \times 10^{-20} \, m^2}$$

$$= 0.068 \, J/m^2$$

$$= 68 \, mN/m$$

The measured value of γ of CCl_4 is 27 mN/m (Table 2.1). The large difference between these two quantities can be ascribed to the idealized assumption that a Stefan ratio of 2 was used in this example.

2.9 INTERFACIAL TENSION OF LIQUID$_1$ (OIL)–LIQUID$_2$ (WATER)

It is a well-known adage that oil and water do not mix. However, it is known that by changing the interfacial forces at the oil–water boundary, one can disperse oil in water (or vice versa). At the oil–water interface, there exists IFT, which can be measured by some of the methods mentioned in Chapter 2 (e.g., drop weight, pendant drop, or Wilhelmy plate). Another well-known phrase is "like molecules like each other" (oil molecules like oil, while polar molecules like polar molecules) (Tanford, 1980).

The IFT, γ_{AB}, between two liquids with γ_A and γ_B is of interest in systems such as emulsions and wetting (Adamson and Gast, 1997; Chattoraj and Birdi, 1984; Somasundaran, 2006). An empirical relation has been suggested (Antonow's rule) by which one can predict the surface tension γ_{AB}:

$$\gamma_{AB} = \left| \gamma_{A(B)} \gamma_{B(A)} \right| \tag{2.55}$$

The prediction of γ_{AB} from this rule is approximate but has been found to be useful in a large number of systems (such as alkanes: water), with some exceptions (such as water: butanol) (Table 2.3). For example,

$$\gamma_{water} = 72 \, mN/m \left(\text{at } 25°C\right)$$

$$\gamma_{hexadecane} = 20 \, mN/m \left(\text{at } 25°C\right)$$

$$\gamma_{water-hexadecane} = 72 - 20 = 52 mN/m (\text{measured} = 50 mN/m) \tag{2.56}$$

TABLE 2.3

Antonow's Rule and Interfacial Tension Data (mN/m)

Oil phase	w(o) = water phase saturated with oil	o(w) = oil phase saturated with water	o/w = at equilibrium	(w(o) (−o = instant))
Benzene	62	28	34	34
Chloroform	52	27	23	24
Ether	27	17	8	9
Toluene	64	28	36	36
n-Propylbenzene	68	29	39	40
n-butylbenzene	69	29	41	40
Nitrobenzene	68	43	25	25
i-Pentanol	28	25	5	3
n-Heptanol	29	27	8	2
CS$_2$	72	52	41	20
Methyleneiodide	72	51	46	22

However, for general consideration, and when exact data is not available, one can use it as a reliable guideline. This model was based on very simple assumptions. The Antonow rule can be understood in terms of a simple physical model. There should be an adsorbed film or Gibbs monolayer of substance B (the substance with lower surface tension) on the surface of liquid A. If one regards this film as having the properties of bulk liquid B, then γ A(B) is effectively the IFT of a duplex surface, and will be equal to $[\gamma_{A(B)} + \gamma_{B(A)}]$.

2.9.1 MEASUREMENT OF IFT BETWEEN TWO IMMISCIBLE LIQUIDS

IFT can be measured by different methods, depending on the characteristics of the system:

* The Wilhelmy plate method
* The drop weight method (can also be used for high pressure and temperature)
* The drop shape method (can also be used for high pressure and temperature)

The Wilhelmy (platinum) plate is placed on the surface of the water, and the oil phase is added till it covers the whole plate. The apparatus must be calibrated with known IFT data, such as water–hexadecane (52 mN/m; 25°C) (Table 2.4).

The drop weight method is carried out by using a pump to deliver the water phase into the oil phase (or vice versa, as one finds suitable). The water drops sink to the bottom of the oil phase. The weight of drops is measured (by using an electro-balance), and IFT can be calculated. The accuracy can be very high if the right kind of setup is chosen.

TABLE 2.4

Interfacial Tensions (IFT) between Water and Organic Liquids (20°C)

Water/Organic Liquid	IFT (mN/m)
n-Hexane	51.0
n-Octane	50.8
CS_2	48.0
CCl_4	45.1
BrC_6H_5	38.1
C_6H_6	35.0
$NO_2C_6H_{5C}$	26.0
Ethyl ether	10.7
n-Decanol	10
n-Octanol	8.5
n-Hexanol	6.8
Aniline	5.9
n-Pentanol	4.4
Ethyl acetate	2.9
Iso-butanol	2.1
n-Butanol	1.6

The drop shape (pendant drop) is most convenient if small amounts of fluids are available, or if extreme temperature and pressures are involved. Modern digital image analyses also make this method very easy to apply in extreme situations. A high pressure and temperature cell with sapphire windows (1 cm thick; can operate at up to 2000 atm pressure) can be used for observation.

3 Surface Active and Fracture-Forming Substances (Soaps and Detergents, etc.)

3.1 INTRODUCTION

In the case of the flow of a fluid (such as water or oil) in porous media (such as oil/gas reservoirs, shale reservoirs, groundwater flow, or similar phenomena), one has *interfacial forces* (liquid–solid interface). For example, higher pressures are needed to recover oil from rock with low permeability than from rock with high permeability, due to capillary pressure (Equation 2.26) (Cernica, 1982; Engelder et al., 2014; Birdi, 2016). The interfaces involved are

- Liquid (water, oil) phase
- Solid (shale, etc.) phase

It is obvious that in all kinds of systems where water is the main fluid, the effect of additives on the surface tension (and related properties) will be significant. In the case of *hydraulic fracking*, fluid (water or another liquid) at high pressure is used to achieve certain well-defined effects for oil or gas recovery (Cahoy et al., 2013; Engelder et al., 2014) (Appendix I). Based on fundamental studies, different additives are used to achieve the specific characteristics needed for the system. The surface forces, which are considered in such a system, are

- Liquid–solid
- Solid–solid (fracture process)

Furthermore, the change in status of fluid in contact with a solid will also contribute to interfacial phenomena. In fracking, the fluid is first pushed into the reservoir under pressure. After the fracking has initiated, the desorbed gas pushes the fluids (both the fracking fluid plus the salt water in the reservoir) up to the borehole. The *wetting* of the reservoir (Chapter 4) will thus determine the characteristics of this operation. Since large volumes of fluids are involved (Engelder et al., 2014; Pagels et al., 2011; Striolo et al., 2012), the capillary forces and the surface wetting energy need to be investigated.

Experiments show that any physical property of a liquid will change when a substance (called *solute*) is dissolved in it. It is found that the change may be small or large, depending on the concentration and other parameters. Experiments have also shown that the magnitude of the surface tension of a liquid will change when a solute is dissolved in it (Chattoraj and Birdi, 1984; Defay et al., 1966; Adamson and Gast, 1997; Birdi, 2014, 2016; Somasundaran, 2015). Also, if one could manipulate the surface tension of water (for example) then many applications areas where surface tension is the important characteristic would be greatly affected.

It is found that there are some specific substances that are used to change (i.e., decrease) the surface tension of water in order to apply this characteristic for some useful purpose in everyday life. As already delineated, fracking water contains substances that are known to assist in fracture formation (*surface-active fracture substances* [SAFS], etc.) (Rehbinder and Schukin, 1972). Experiments show that the magnitude and characteristic of surface tension change will depend on the solute added and on its concentration (Chattoraj and Birdi, 1984; Adamson and Gast, 1997; Defay et al., 1966; Scheludko, 1966; Schramm, 2010; Somasundaran, 2015; Birdi, 2016). For example, in some cases, the surface tension of the water solution increases (with the addition of NaCl, for example), while, in other cases, it decreases with the addition of surface-active substances (SAS; such as ethanol, methanol, soaps, etc.). The change in surface tension, γ, may be small (per mole added) (as in the case of inorganic salts), or large (as in the case of such molecules as ethanol or other soap-like molecules).

Change in γ with the addition of solute (as gram per liter):

- Inorganic salt: *Minor change* (increase) in γ
- Ethanol, or similar: *Small change* (decrease) in γ
- Soap, or similar): *Large change* (decrease) in γ

The following is some typical surface tension data of different solutions:

Surface tension (γ) (mN/m)	72	50	40	30	22
Surfactant ($C_{12}H_{25}SO_4Na$)	0	0.0008	0.003	0.008	
Ethanol	0	10	20	40	100

This shows that to reduce the value of γ of water from 72 to 30 mN/m, one would need 0.005 moles of sodium dodecyl sulfate (SDS) or 40% ethanol. Of course, these two solutions cannot be used for the same application based on their similar magnitude of γ. SDS solutions exhibit other characteristics, which makes them different from an ethanol solution. There are special substances called *soaps* or *detergents* or *surfactants* ($C_{12}H_{25}SO_4Na$: SDS), which exhibit unique physico-chemical properties. The most significant structure of these molecules is due to the presence of a *hydrophobic* (alkyl) group and a *hydrophilic* (polar) group (such as OH, $-CH_2CH_2O-$, $-COONa$, $-SO_3Na$, $-SO_4Na$, $-CH_33N-$, etc.).

The different polar groups are
Ionic groups

- (negatively charged: *Anionic*)
- $-COONa$
- $-SO_3Na$
- $-SO_4Na$
- $-(N)(CH_3)_4Br$ (positively charged: *Cationic*)
- $-(N)(CH_3)_2-CH_2-COONa$ (Amphoteric)

Nonionic groups

- $-CH_2CH_2OCH_2CH_2OCH_2CH_2OH$
- $-(CH_2CH_2OCH_2CH_2O)_x(CH_2CH_2CH_2O)_yO$
- Accordingly, one also calls these substances *amphiphilic* (also *amphipathic*), meaning having both hydrophilic and hydrophobic parts, i.e., alkyl and polar.
- CCCCCCCCCCCCCCCCCCCCC-**O**
- Alkyl group (CCCCCCC-)–polar group (**–O**) = amphiphile
- $(CH_3CH_2CH_2CH_2CH_2CH_2CH_2CH_2CH_2)$-polar

For instance, surfactants dissolve in water and give rise to low surface tension (even at very low concentrations [1–100 mmol/L]) of the solution. Therefore, these substances are also called *surface-active molecules* (*surface-active agents* or substances). On the other hand, most inorganic salts increase the surface tension of water. All surfactant molecules are amphiphilic, which means that these molecules exhibit both hydrophilic and hydrophobic properties. Ethanol reduces the surface tension of water, but one will need more than a few moles per liter to obtain the same reduction as when using a few moles of surface-active agents. In comparison, the addition of organic molecules such as methanol or ethanol decreases the magnitude of γ of water from 72 mN/m rather slowly. The value of γ decreases from 72 to 22 mN/m in pure ethanol. In comparison, the value of γ of surfactant solutions decreases to 30 mN/m with a surfactant concentration around millimoles per liter (range of 1–10 g/L). Soaps have been used by mankind for many centuries. In biology, one finds a whole range of amphiphile molecules (bile salts, fatty acids, cholesterol, and other related molecules, known as *phospholipids*).

It is important to mention that surfactants are one of the most important types of substances, which play an essential role in everyday life. Many surfactants exist in nature (such as bile acids in the stomach, which behave exactly the same way as man-made surface-active agents). Proteins, which are large molecules with molecular weights from 6000 to over a million, also decrease the value of γ when dissolved in water.

Surfactants are characterized as amphiphiles. Amphiphile is a Greek word that means likes both kinds. A part of the amphiphile likes oil and is hydrophobic (or lipophilic), while the other part likes water and is hydrophilic (or lipo-phobic). The balance between these two parts, hydrophilic–lipophilic, is called the *hydrophilic–hydrophobic balance* (HLB). The latter quantity can be estimated by experimental means, and theoretical analyses allow one to estimate its value (Adamson and Gast, 1997; Holmberg, 2002; Hansen, 2000; Birdi, 2007, 2016). HLB values are applied in the emulsion industry. Soap molecules are made by reacting fats with strong alkaline solutions (this process is called *saponification*). In water solution, the soap molecule $C_nH_{2n+1}COONa$ (with n greater than 12–22), dissociates at high pH into $RCOO^-$ and Na^+ ions.

The major characteristic, besides the lowering of γ by SAS, is the effect of ionic charge (zero or negative or positive) on the surface. The latter property means that the interface of the water solution of an SAS may be

- Zero charge
- Negative charge
- Positive charge

In many applications, one finds it necessary to employ surfactants that are non-ionic. Nonionic surfactants of many kinds have been synthesized, and one can obtain tailor-made surfactants to suit a particular application. Also, since nonionic detergents do not exhibit any charge, these find applications where this property is essential.

3.2 SURFACE TENSION OF AQUEOUS SOLUTIONS (GENERAL REMARKS)

The surface tension γ of any pure liquid (water or organic liquid) will change when another substance (solute) is dissolved. The change in γ will depend on the concentration and the characteristics of the solute added. The surface tension of water is found to increase (in general) when inorganic salts (such as NaCl, KCl, or Na_2SO_4) are added (Figure 3.1), while its value decreases when organic substances (ethanol, methanol, fatty acids, soaps, detergents) are dissolved (Figure 3.1).

The surface tension of water at 20°C *increases* from 72 mN/m (dyne/cm) to 73 mN/m when 1 M (59 g/L) NaCl is added. On the other hand, the magnitude of the surface tension *decreases* from 72 to 39 mN/m when only 0.008 M (0.008 M × 288 = 2.3 g/L) SDS (mol. wt. = 288) is dissolved.

For example, the surface tension data of short-chain alcohol (n-butanol) solutions in water are found to decrease from 72 mN/m (pure water) to 50 mN/m in 200 mmol/L (Figure 3.1). The magnitude of γ of water changes slowly on addition of methanol as compared with detergent solutions. The methanol–water mixtures gave the following γ data (at 20°C).

%w methanol	0	10	25	50	80	90	100
γ		72	59	46	35	27	25	22.7

The surface tension data in the case of a homologous series of alcohols and acids show some simple relation to the alkyl chain length (Figure 3.2). It is noticed that each addition of $-CH_2-$ group in the alkyl chain, gives a value of surface tension such that the value of the concentration is lower by about a factor 3.

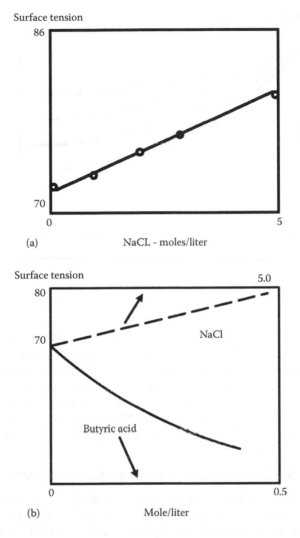

FIGURE 3.1 Change in surface tension of water as a function of added solutes (inorganic salts, surface-active agents) (NaCl, butyric acid): (a) NaCl solutions.

However, experiments show that such dependence in the case of nonlinear alkyl chains will be different (Birdi, 1997, 2010b, 2016). The effective $-CH_2-$ increase in the case of nonlinear chains will be less (ca. 50%) than in the case of a linear alkyl chain. The tertiary $-CH_2-$ group effect would be even less. In general, though, one will expect that the change in γ per mole of substance will increase with any increase in the hydrocarbon group of the amphiphile.

The effect of chain length on surface tension arises from the fact that as the hydrophobicity increases with each $-CH_2-$ group, the amphiphile molecule adsorbs more at the surface (Appendix III). This will thus also be a general trend in more complicated molecules, such as in proteins and other polymers. In proteins, the amphiphilic

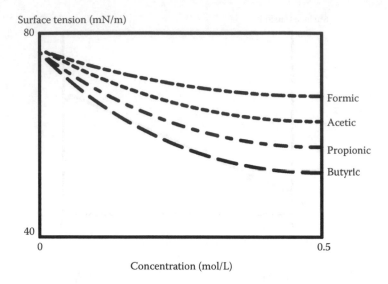

FIGURE 3.2   Surface tension data of some homologous series of short-chain acids in water.

property arises from the different kinds of amino acids (25 different amino acids). Some amino acids have lipophilic groups (such as phenylalanine, valine, leucine, etc.), while others have hydrophilic groups (such as glycine, aspartic acid, etc.) (Birdi, 2016). In fact, one finds from surface tension measurements that some proteins (such as hemoglobin) are considerably more hydrophobic than others (such as bovine serum albumin [BSA] or ovalbumin). These properties of proteins have been extensively investigated (Tanford, 1980; Chattoraj and Birdi, 1984; Birdi, 2016).

3.2.1   AQUEOUS SOLUTIONS OF SURFACE-ACTIVE SUBSTANCES (SAS) (AMPHIPHILES)

All molecules which when dissolved in water *reduce* surface tension are called *surface-active substances* (soaps, surfactants, detergents, alcohols, proteins). This means that SAS adsorb (with high affinity) at the surface and reduce surface tension. The same will happen if one adds an SAS to an oil–water system. The interfacial tension of the oil–water interface will be reduced accordingly. Inorganic salts, on the other hand, increase the surface tension of water (with a few exceptions, such as urea).

Surfactants (soaps, etc.), exhibit *surface activity*, which means their molecules will adsorb preferentially at the following interfaces:

- Air–water
- Oil–water
- Solid–water

The magnitude of the surface tension is reduced since the hydrophobic (alkyl chain or group) is energetically more attracted to the surface than being surrounded

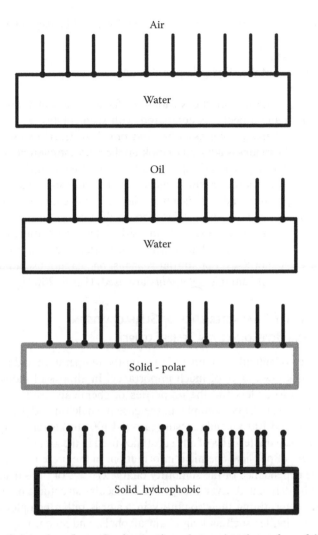

FIGURE 3.3 Orientation of soap (surface-active substance) at the surface of different interfaces: air–water, oil–water, water–solid (polar), water–solid (hydrophobic).

by water molecules inside the bulk aqueous phase. Figure 3.3 shows the monolayer formation of the SAS at high-bulk concentration. Since the close-packed SAS at the surface looks like alkane, it would be thus expected that the surface tension of the SAS solution would decrease from 72 mN/m (surface tension of pure water) to a more alkane-like surface tension (close to 25 mN/m).

The orientation of the surface molecule at the interface will be dependent on the system. This is shown as follows:

- *Air–water*: Polar part toward water and hydrocarbon part toward air
- *Oil–water*: Polar part toward water and hydrocarbon part toward oil
- *Solid–water*: Polar part toward water and hydrocarbon part toward solid

In other words, the surface properties of any SAS will be dependent on

- The decrease in γ
- The structure of the alkyl group of the SAS

The solution properties of the various surfactants in water are very unique and complex in many aspects, as compared with such solutes as NaCl or ethanol. However, in this chapter, some essential and practical descriptions will be given. For more detail, the reader is advised to look up the relevant references (Birdi, 2002, 2016; Tanford, 1980; Rosen and Kunjappu, 2012; Somasundaran, 2015). The solubility of charged and noncharged surfactants is very different, especially with regard to the effect of temperature and salts (such as NaCl). These characteristics are important when one needs to apply these substances in diverse systems. For instance, one cannot use the same soap in seawater as in fresh water. The main reason being that salts (such as Ca^{++} and Mg^{++}), as found in seawater, affect the foaming and solubility characteristics of major SAS. For similar reasons, one cannot use a nonionic detergent for shampoos (only anionic detergents are used) (Birdi, 2003a).

3.2.2 Solubility Characteristics of Surfactants in Water (Dependence on Temperature)

The dependence of solubility characteristics on the temperature and pressure of the surroundings is known to be of much importance. In almost all industrial applications, one needs to understand the properties of chemicals used at different temperatures and pressures. For example, in the case of shale oil and gas reservoirs, the temperature (around 100°C) and pressure (around 100 atm) is relatively very high.

The solubility characteristics of any substance used in these environments are very important pieces of physical chemical information. In the present case, the information about the dependence of the solubility characteristics of the surfactant on temperature will be delineated. Even though the molecular structures of surfactants are rather simple, determining their solubility in water is rather complex as compared with other amphiphiles such as long chain alcohols, and so on. Experiments show that solubility in water is dependent on the alkyl chain length. However, it is also found that the solubility of surfactants is dependent on the presence of a charge on the polar group. The ionic surfactants exhibit different solubility characteristics than the nonionic surfactants, with regard to dependence on the temperature. In fact, in all industrial applications of SAS, the solubility parameter is one of the most important. This characteristic is the determining factor when deciding which SAS is to be used in a given system. For example, the SAS one will need in household washing detergent will be different from the SAS one will need when seawater is used for washing, and so on. The hydrophobic alkyl part exhibits solubility in water, which has been related to a surface tension model of the *cavity* (Appendix IV).

3.2.2.1 Ionic Surfactants

The solubility of all ionic surfactants (both anionics, which are negatively charged, and cationics, which are positively charged) is low at low temperature but at a specific

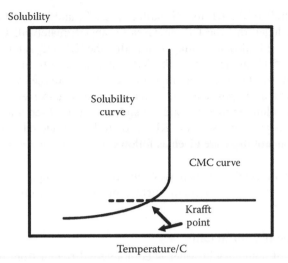

FIGURE 3.4   Solubility (Kraft point: KP) of ionic (anionic or cationic) surfactants in water (as a function of temperature).

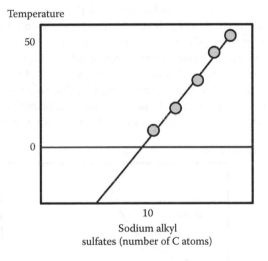

FIGURE 3.5   Variation of KP with chain length of sodium alkyl sulfates.

temperature the solubility suddenly increases (Figure 3.4). For instance, the solubility of SDS at 15°C is about 2 g/L. This temperature is called the *Krafft point* (KP). The solubility of SDS increases drastically above its KP. A KP can be obtained by cooling an anionic surfactant solution (ca. 0.5 molar) from a high to a lower temperature until cloudiness suddenly appears. The KP is not very sharp in the case of impure surfactants, as generally found in industry.

In fact, the solubility near the KP is almost equal to the CMC. The magnitude of the KP is found to be dependent on the chain length of the alkyl chain (Figure 3.5).

The magnitude of the KP for Cl_2 sulfate is 21°C, and it is 34°C for Cl_4 sulfate. It may be concluded that the KP increases by approximately 10°C per CH_2 group (linear relation). It is also interesting to note that the KP of C_8 is found from extrapolation to be −3.5°C. In fact, for C_8 the KP is not possible to measure from experiments. The industrial products are mixtures and therefore the magnitude of the KP is dependent on the composition. Since no *micelles* can be formed below the KP, this means that the solution properties are dependent on this observation. Therefore, the effect of various parameters on the KP needs to be considered in the case of ionic surfactants. Some of these are given as follows:

- Alkyl chain length (KP increases with alkyl chain length).
- KP decreases if lower-chain surfactant is mixed with a longer chain surfactant.

3.2.2.2 Nonionic Surfactants

The solubility of nonionics in water is completely different from those of charged surfactants (especially with regard to the effect of temperature). The solubility of nonionic surfactants is high at low temperature but it decreases abruptly at a specific temperature, called the *Cloud point* (CP) (Figure 3.6). This means that nonionic detergents will not be suitable if used *above* the CP. The solubility of such detergent molecules in water arises from the hydrogen bond formation between the hydroxyl (−OH) and ethoxy groups (-CH_2CH_2O-) and the water molecules. At high temperatures, the degree of hydrogen bonding gets weaker (due to high molecular vibrations) and thus the nonionic detergents become insoluble at the CP. The CP is named for the temperature at which the solution becomes cloudy. The solution separates into two phases, with a rich water phase and low concentration of nonionic surfactant. The rich nonionic detergent phase is found to consist of low water content. Experiments

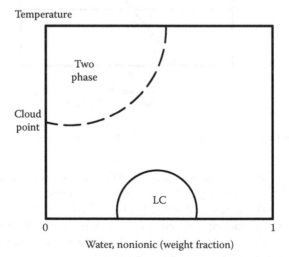

FIGURE 3.6 Solubility of a nonionic surfactant in water (Cloud-point: CP) (dependent on temperature).

have shown that there are roughly four molecules of bound water per ethylene oxide group ($-CH_2CH_2O-$) (Birdi, 2007).

Furthermore, the charge of a surfactant is a very characteristic property, with regard to its application. Anionic surfactants are used in completely different applications to cationic surfactants. For instance, anionic surfactants are used for shampoos and washing, while cationics are used for hair conditioners. Hair has a negative ($-$) charged surface and thus the cationics strongly adsorb at the surface and leave a smooth surface. This is because the charged end is oriented toward the hair surface and the alkyl group is pointing away (as depicted next).

- Cationic detergent + hair
- Alkyl group: polar group(+)hair($-$)

Experiments show that similar adsorption behavior is observed on all kinds of charged solid surfaces (e.g., minerals, glass, metals, paper and pulp, and plastics). For example, cationic detergents are used to reduce attraction between silica (with negative charges) surfaces (as used in insulation materials).

3.3 MICELLE FORMATION OF SURFACTANTS (IN AQUEOUS MEDIA)

Surfactant aqueous solutions manifest two major forces, which determine the solution behavior. The alkyl part being hydrophobic tends to separate out as a distinct phase, while the polar part tends to stay in the aqueous solution. The surface tension of a surfactant solution is shown in Figure 3.7.

It is seen that the magnitude of the surface tension decreases until, at a specific critical micelle concentration (CMC), it abruptly stops any further decrease. Similar changes are observed if one measures other physical characteristics of surfactant solutions (such as conductivity, density, foaming, bubble formation, cleaning and detergency effectiveness, etc.). The difference between these two opposing forces thus determines the solution properties. The factors that one has to consider are the following:

a. The alky group and water
b. The interaction of the alkyl hydrocarbon groups with themselves
c. The solvation (through hydrogen bonding and hydration with water) of the polar groups
d. Interactions between the solvated polar groups

Below CMC, the detergent molecules are present as single monomers. One may even say that below CMC the substance behaves as normal. Above CMC, monomers, C_{mono}, are in equilibrium with micelles, C_{mice}. A micelle, with its aggregation number, N_{ag}, is formed from monomers (Figure 3.7):

$$N_{ag} \text{ Monomer} = \text{Micelle} \qquad (3.1)$$

N_{ag} monomers, which were surrounded by water, aggregate together, above CMC, and form a micelle. In this process, the alkyl chains have transferred from the water

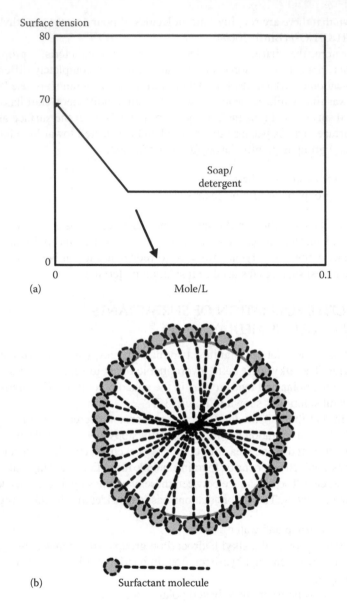

(a)

(b)   Surfactant molecule

FIGURE 3.7   (a) Change of surface tension of a typical detergent solution with concentration; (b) a typical micelle with spherical shape (the diameter is almost equal to twice the length of the molecule).

phase to an alkane-like micelle interior. This occurs because the alkyl part is at a lower energy in the micelle than in the water phase (as shown next in case a):

a. Alkyl chain in water/surrounded by water
b. Micelle: alkyl chain in contact with neighboring alkyl chains

Thus, in case b, the repulsion between alkyl chain and water will be absent. Instead, the alkyl–alkyl attraction in case b is the driving force for the micelle formation. The surfactant molecule forms a micellar aggregate at a concentration higher than CMC, because it moves from the water phase to the micelle phase (lower energy). The micelle reaches equilibrium after a specific number of monomers have formed a micelle (as determined by the free energy of the system). This means that there are both attractive and opposing forces involved in this process. Otherwise, one would expect very large aggregates if there was only attractive forces involved. At equilibrium, there must exist a state where two different forces are equal. Thus, one can write the standard free energy of a micelle formation, ΔG°_{mice}, as

$$\Delta G^{\circ}_{mice} = \text{attractive forces} + \text{opposing forces} \tag{3.2}$$

The attractive forces are associated with the hydrophobic interactions between the alkyl part (alkyl–alkyl chain attraction) of the surfactant molecule, $\Delta G_{hydrophobic}$. The opposing forces arise from the polar part (charge–charge repulsion, polar group hydration), ΔG_{polar}. These forces are of opposite signs. The attractive forces would lead to larger aggregates. The opposing forces would hinder the aggregation. A micelle with a definite aggregation number is where the value of ΔG°_{mice} is zero. Hence, we can write for ΔG°_{mice}

$$\Delta G^{\circ}_{mice} = \Delta G^{\circ}_{hydrophobic} + \Delta G^{\circ}_{polar} \tag{3.3}$$

The standard free energy of micelle formation will be

$$\Delta G^{\circ}_{mice} = \mu^{\circ}_{mice} - \mu^{\circ}_{mono}$$

$$= RT \ln\left(\frac{C_{mice}}{C_{mono}}\right) \tag{3.4}$$

At CMC, one may neglect C_{mice}, which leads to

$$\Delta G^{\circ}_{mice} = RT \ln\left(CMC\right) \tag{3.5}$$

This relation holds for nonionic surfactants, but will be modified in the case of ionic surfactants (Equation 3.6). This equilibrium shows that if we dilute the system, then micelles will break down to monomers to achieve equilibrium. This is a simple equilibrium for a nonionic surfactant. In the case of ionic surfactants, there will be charged species. For example, the ionic surfactant aqueous solution of SDS, the micelle with aggregation number, N_{SD-}, will consist of counter-ions, C_{S+}:

$$N_{SD-} \text{ ionic surfactant monomers} + C_{S+} \text{ counter} = \text{micelle with charge} \left(N_{SD-} - C_{S+}\right)$$

$$\tag{3.6}$$

Since N will be larger than S^+, all anionic surfactants are negatively charged. Similarly, all cationic micelles will be positively charged. For instance, for cetyl trimethyl ammonium bromide (CTAB), we have the following equilibria in micellar solutions:

$$CTAB \rightleftharpoons dissociates\ into\ CTA^+\ and\ Br^-\ ions$$

The micelle with N_{CTA+} monomers will have C_{Br-} counter-ions. The positive charge of the micelle will be the sum of positive and negative ions ($N_{CTA+} - C_{Br-}$). The actual concentration will vary in each species with the total detergent concentration (as in the case of SDS solutions, Figure 3.8).

Below CMC, the SDS molecules in water are found to dissociate into SD^- and Na^+ ions. Conductivity measurements show:

a. That SDS behaves as a strong salt and SD^- and Na^+ ions are formed (the same as one observes for NaCl).
b. That a break on the plot is observed at SDS concentration equal to the CMC. This clearly shows that the number of ions decreases with concentration. The latter indicates that some ions (in the present case, cations, Na^+) are partially bound to the SDS micelles, which results in a change in the slope of the conductivity of the solution. The same behavior is observed in other ionic detergents, such as cationic (CTAB) surfactants.

At CMC, micelles (aggregates of SD^- with some counter-ions, Na^+) are formed and some Na^+ ions are bound to these, which is also observed from conductivity data. In fact, these data analyses have shown that approximately 70% of Na^+ ions are

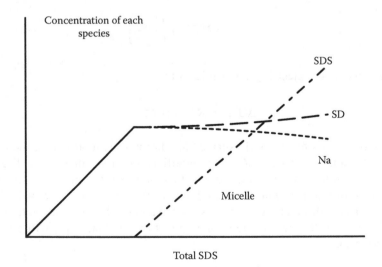

FIGURE 3.8 Variation of concentration of different ionic species for SDS solutions (Na^+, SD^-, $SDS_{micelle}$).

bound to SD⁻ ions in the micelle. The surface charge was estimated from conductivity measurements (Tanford, 1980; Birdi, 2002; Somasundaran, 2015). Therefore, the concentration of Na^+ will be higher than SD⁻ ions after CMC. A large number of reports are found in the literature where the transition from the monomer phase (before CMC) to the micellar phase (after CMC) has been analyzed. The same is true in the case of cationic surfactants. In the case of CTAB solutions, one thus has CTA^+ and Br^- ions below CMC. Above CMC, there are additionally CTAB micelles. In these systems, the counter-ion is Br^-.

- Analyses of CMC: The CMC will be dependent on different factors. These will be dependent on both the alkyl part and the polar part. The interaction of the detergent with the solvent will also have an effect on the CMC.
- Effect of alkyl chain length: It has been found that the CMC decreases with increasing alkyl chain length.

For example, the following relation has been found for Na-alkyl sulfate detergents:

$$\ln\left(CMC\right) = k_1 - k_2\left(C_{alkyl}\right) \tag{3.7}$$

where:

k_1 and k_2 are constants
C_{alkyl} is the number of carbon atoms in the alkyl chain

The CMC will change if the additive has an effect on the monomer–micelle equilibrium. It will also change if the additive changes the detergent solubility. The CMC of all ionic surfactants will decrease if co-ions are added. However, nonionic surfactants show very little change in CMC on the addition of salts. This is as one should expect from theoretical considerations. The change of CMC (at 25°C) with NaCl for SDS solutions is as follows (Figure 3.9):

NaCl (mole/L)	CMC (mol/L)	g/L	N_{agg}
0	0.008	2.3	80
0.01	0.005	1.5	90
0.03	0.003	0.09	100
0.05	0.0023	0.08	104
0.1	0.0015	0.05	110
0.2	0.001	0.02	120
0.4	0.0006	0.015	125

The most important feature, which must be noticed in the data, is that very small amounts of electrolytes have very strong effects on the system. This effect is the same for all charged (negative or positive) surfactant molecules. The radius of the spherical micelle is reported as 20 Å, which increases to 23 Å (for the nonspherical) (Figure 3.9).

These data convincingly show that the ionic interactions at the micelle interface are strongly dependent on the surrounding ions. Experiments have shown that, in

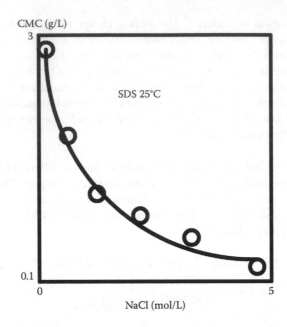

FIGURE 3.9 Variation of CMC of SDS solutions with added NaCl for micelles.

most cases, such as for SDS solutions, the initial spherical micelles may grow under some influence into larger aggregates (from spherical to ellipse), and that very large micelles are found to take the shape of discs or cylinders or lamellas (Figure 3.10).

It is important to notice how CMC changes with even a very small addition of NaCl (electrolyte). It has been found in general that the change in CMC with the addition of ions follows the relation:

$$\ln\left(\text{CMC}\right) = \text{Constant}_1 - \text{Constant}_2\left(\text{Ln}\left(\text{CMC} + \text{C}_{\text{ion}}\right)\right) \qquad (3.8)$$

The quantity Constant_2 was found to be related to the *degree of micelle charge*. Its magnitude varied from 0.6 to 0.7, which means that micelles have a 30% charge. Data show that the CMC of cationic surfactants decreased on the addition of KBr as follows

- DTAB (dodecyltrimethylammonium bromide), TrTAB (tridecyl tri-methyl ammonium bromide), and TTAB (tetradecyl trimethyl ammonium bromide):
- $\text{Ln}\left(\text{CMC}\right) = -6.85 - 0.64 \, \text{Ln} \, \left(\text{CMC} + \text{C}_{\text{KBr}}\right)$
- $\text{TrTAB}: \text{Ln}\left(\text{cmc}\right) = -8.10 - 0.65 \, \text{Ln}\left(\text{CMC} + \text{CKBr}\right)$
- $\text{TTAB}: \text{Ln}\left(\text{cmc}\right) = -9.43 - 0.68 \, \text{Ln}\left(\text{CMC} + \text{CKBr}\right)$

It is noticed that the slope increases with the increase in alkyl chain length. A similar relationship has been reported for an Na-alkyl sulfate homologous series.

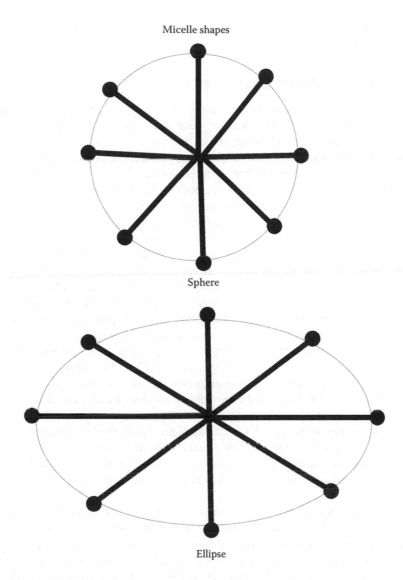

Micelle shapes

Sphere

Ellipse

FIGURE 3.10 Different types of micellar aggregates: (a) spherical; (b) disc-like; (c) cylindrical.

The CMC data for soaps are found to give the following dependence on the alkyl chain length:

Soap	CMC (mole/L) 25°C
C_7COOK	0.4
C_9COOK	0.1
$C_{11}OOK$	0.025

These data show that CMC decreases by a factor of *4 for each increase in chain length* by $-CH_2CH_2-$.

3.4 GIBBS ADSORPTION EQUATION IN SOLUTIONS

A pure liquid, (such as water) when shaken, does not form any foam. This merely indicates that the surface layer consists of pure liquid (and the absence of any minor surface-active impurities). However, if one adds a very small amount of surface-active agent (soap or detergent, ca. millimole concentration: or about ppm by weight) and shakes the solution, foam is formed at the surface. In nature, one finds a variety of SAS that cause foam formation (as one notices on the shores of lakes, rivers, and oceans). This indicates that the surface-active agent has *accumulated* at the surface (meaning that the concentration of surface-active agent is much higher at the surface than in the bulk phase: in some cases many thousand times more concentrated) and thus forms a thin liquid film (TLF), which constitutes the bubble. In fact, one can use the bubble or foam formation as a useful measure of the purity of the system. One generally observes at the shores of lakes or oceans that foam bubbles are formed under different conditions. If water in these sites is polluted, then very stable foams are observed. If one adds instead an inorganic salt, NaCl, then no foam is formed. The foam formation indicates that the surface-active agent adsorbs at the surface and forms a TLF (consisting of two layers of amphiphile molecule and with some water), as found from analyses (Scheludko, 1966; Birdi, 2016). This has led to many theoretical analyses of surfactant concentration (in the bulk phase) and the surface tension (which will be related to the presence of surfactant molecules at the surface). The thermodynamics of surface adsorption has been extensively described by Gibbs adsorption theory (Defay et al., 1966; Chattoraj and Birdi, 1984). Furthermore, Gibbs adsorption theory is found to be fundamental for other systems than solutions, such as solid–liquid or $liquid_1$–$liquid_2$, adsorption of solute on polymers, etc. In fact, in any system where adsorption takes place at an interface, the Gibbs theory will be applicable.

The addition of a solute to water leads to a change of composition in the bulk phase, and also at the surface. In other words, both the bulk and the surface properties change. For example, the surface tension, γ, of water changes with the addition of organic or inorganic solutes, at constant temperature and pressure (Defay et al., 1966; Chattoraj and Birdi, 1984; Birdi, 1989, 1997, 2016; Somasundaran, 2015). The extent of surface tension change and the sign of change are determined by the molecules involved (see Figure 3.1).

The magnitude of γ of aqueous solutions generally increases with different electrolyte concentration. The magnitude of γ of aqueous solutions containing organic solutes invariably decreases. As mentioned in Chapter 2, the surface of a liquid is where the density of the liquid changes to that of a gas, by a factor of 1000 (Figure 3.1). Now, let us look at what happens to surface composition when ethanol is added to water (Figure 3.11a,b).

The reason ethanol concentration in the vapor phase is higher than water is due to its lower boiling point. This phenomena is common in a cognac glass where ethanol vapors are noticed on the inner side. Next, let us consider the situation when

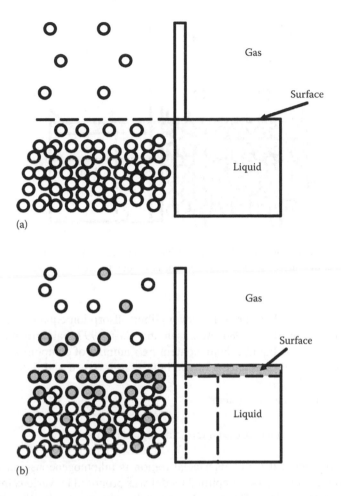

FIGURE 3.11 Surface composition: (a) of pure water; (b) of an ethanol–water solution (shaded = ethanol).

a detergent is added to water, whereby the surface tension is lowered appreciably (Figure 3.12).

The schematic concentration profile of detergent molecules is such that the concentration is homogeneous up to the surface. At the surface, there is almost only detergent molecules plus the necessary number of water molecules (which are in a bound state to the detergent molecule). The surface thus shows very low surface tension, ca. 30 mN/m. The surface concentration profile of a detergent is not easily determined by any direct method. As mentioned in Section 3.2.1, experiments show that the surface tension of water decreases when organic substances such as alcohol (methanol, ethanol, glycerol, glycol) are used—as in *hydraulic fracturing*. This suggests that in general, there are more alcohol molecules at the surface than in the bulk.

The state of the surface composition of water solutions of such systems has been the subject of extensive investigations in the literature. These systems have been

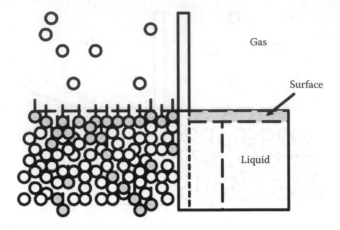

FIGURE 3.12   Concentration of detergent (shaded with tail) in solution and at surface. The shaded area at the surface is the excess concentration due to accumulation.

exhaustively analyzed by the well-known Gibbs adsorption equation (Defay et al., 1966; Chattoraj and Birdi, 1984; Adamson and Gast, 1997; Fainerman et al., 2002; Birdi, 1989, 2016). A liquid column containing i number of components is assumed (according to the Gibbs treatment of two bulk phases, i.e., α and β) to be separated by the *interfacial region*.

Liquid column in a real system:

$$\alpha - \text{phase} \equiv \left(\text{interfacial region}\right) \equiv \beta - \text{phase}.$$

It was considered that this interfacial region is inhomogeneous and difficult to define, and therefore a more simplified model was proposed by various investigators (Adamson and Gast, 1997; Defay et al., 1966; Chattoraj and Birdi, 1984; Birdi, 1997, 2016), in which the interfacial region is assumed to be a mathematical plane. In this actual system, the bulk composition of the ith component in α and ß phase is $c_{i\alpha}$ and $c_{i\beta}$, respectively.

However, in the idealized system, the chemical compositions of the α and ß phases are imagined to remain unchanged right up to the dividing surface so that their concentrations in the two imaginary phases are also $c_{i\alpha}$ and $c_{i\beta}$, respectively (Chattoraj and Birdi, 1984).

If $n_{i\alpha}$ and $n_{i\beta}$ denote the total moles of the ith component in the two phases of the idealized system, then the Gibbs surface excess Γ_{ni} of the ith component can be defined as

$$(n_{i\alpha} \text{ moles in phase } \alpha) - (n_{ix} \text{ moles in surface region as excess})$$
$$-(n_{i\alpha} \text{ moles in phase } n_{i\beta})$$

$$n_i^x = n_i^x - n_{i\alpha} - n_{i\beta} \tag{3.12}$$

where n_i^t is the total mole of the ith component in the real system. In the exact same way, one can define the respective surface excess internal energy, E_x, and entropy, S_x, by the following mathematical relationships (Chattoraj and Birdi, 1984; Birdi, 1989; Somasundaran, 2015):

$$E^x = E^t - E^\alpha - E^\beta \qquad (3.13)$$

$$S^x = S^t - S^\alpha - S^\beta \qquad (3.14)$$

Here, E^t and S^t are the total energy and entropy, respectively, of the system as a whole for the actual liquid system in Figure 3.8.

The energy and entropy terms for α and β phases are denoted by the respective superscripts. The excess (x) quantities thus refer to the surface molecules in the adsorbed state.

The real and idealized systems are open so that the following equation can be written:

$$dE^t = TdS^t - (p\,dV + p'dV' - \gamma dA) + \mu_1\,dn_1{}^t + \mu_2 dn_2{}^t + \cdots + \mu_i dn_i{}^t \qquad (3.15)$$

where:

V^α and V^β are the actual volumes of each bulk phase
p and p'   are the respective pressures

Since the volume of the interfacial region was considered to be negligible, then $V^t = V^\alpha + V^\beta$. Furthermore, if the surface is almost planar, then $p_\alpha = p'_\beta$, and $(p\,dV^\alpha + p_\beta\,dV^\beta) = p\,dV^t$. The changes in the internal energy for idealized phases α and β may similarly be expressed as follows:

$$dE^\alpha = T\,dS^\alpha - p\,dV^\alpha + \mu_1 dn_1 + \cdots + \mu_i dn_{i,\alpha} \qquad (3.16)$$

and

$$dE^\beta = TdS^\beta - p\,dV^\beta + \mu_1 dn_1 + \cdots + \mu_i dn_{i,\beta} \qquad (3.17)$$

In the real system, the contribution due to the change of the surface energy, $\gamma\,dA$, is included as additional work. Such a contribution is absent in the idealized system containing only two bulk phases without the existence of any physical interface. By subtracting Equations 3.16 and 3.17 from Equation 3.15, the following relationship is obtained:

$$d(E^t - E^\alpha - E^\beta) = T\,d(S^t - S^\alpha - S^\beta) + \gamma dA + \mu_1 d(n_t - n_{1\alpha} - n_{1\beta})$$
$$+ \cdots + \mu_i d(n_{i,t} - n_{i\alpha} - n_{i,\beta}) \qquad (3.18)$$

or

$$d\,E^x = T\,dS^x + \gamma dA + \mu_1 dn_1{}^x + \cdots + \mu_i dn_i{}^x \qquad (3.19)$$

This equation on integration at constant T, γ, and μ_i, and so on gives

$$E^x = T\,S^x + \gamma A + \mu_1 n_1{}^x + \cdots + \mu_i n_i{}^x \tag{3.20}$$

This relationship may be differentiated in general to give

$$dE^x = T\,dS^x + \gamma dA +_i \mathrm{SUM}(\mu_i dn_i{}^x) +_i \mathrm{SUM}(n_i d\mu_i{}^x) + A\,d\gamma + S^x dT \tag{3.21}$$

A combination of Equations 3.21 and 3.20, gives

$$-A d\gamma = S^x dT +_i \mathrm{SUM}(n_i{}^x d\mu_i) \tag{3.22}$$

Let $S^{s,x}$ and Γ_{ix} denote the surface excess entropy and moles of the ith component per surface area, respectively. This gives

$$S^{s,x} = \frac{S^x}{A}$$

$$\Gamma_i{}^x = \frac{n_i{}^x}{A} \tag{3.23}$$

and

$$-d\gamma = S^{s,x} dT + \Gamma_{x,1} d\mu_1 + \Gamma_{x,2} d\mu_2 + \cdots + \Gamma_{i,x} d\mu_i \tag{3.24}$$

This equation is similar to the Gibbs-Duhem equations for the bulk liquid system (Chattoraj and Birdi, 1984; Birdi, 2016). The relation in Equation 3.24 can also be rewritten as

$$-d\gamma = S_{s,1} dT + \Gamma_2{}^1 d\mu_2 + \cdots + \Gamma_i{}^1 d\mu_i \tag{3.25}$$

$$= S_{s,1} dT + \Gamma_i{}^1 d\mu_i \tag{3.26}$$

At constant **T** and **p**, for a two-component system (say water(1) + alcohol(2)), we thus obtain the classical Gibbs adsorption equation as

$$\Gamma_2 = -\left(\frac{d\gamma}{d\mu_2}\right)_{T,p} \tag{3.27}$$

This shows that the quantity, Γ_2, is related to the *change* in γ with the change in the chemical potential of the additive (solute). The chemical potential μ_2 is related to the activity of alcohol by the equation:

$$\mu_2 = \mu_2^{\circ} + RT \ln(a_2) \tag{3.28}$$

If the activity coefficient can be assumed to be equal to unity, then

$$\mu_2 = \mu_2^\circ + RT \ln(C_2) \tag{3.29}$$

where C_2 is the bulk concentration of solute 2. The Gibbs adsorption then can be written as

$$\Gamma_2 = -\frac{1}{RT}\left(\frac{d\gamma}{d\ln(C_2)}\right)$$

$$= -\frac{C_2}{RT}\left(\frac{d\gamma}{dC_2}\right) \tag{3.30}$$

This shows that the *surface excess* quantity on the left-hand side is proportional to the change in surface tension with the concentration of the solute ($d\,\gamma/d(\ln(C_{surfaceactivesubstance})))$. A plot of $\ln(C_2)$ versus γ gives a slope equal to

$$\Gamma_2(RT)$$

from this, one can estimate the value of Γ_2 (moles/area).

This shows that all SAS, will always have a higher concentration at the surface than in the bulk of the solution.

This relation has been verified by using radioactive tracers. Also, as will be shown later under spread monolayers, one finds a very convincing support to this relation and the magnitudes of Γ for various systems. The surface tension of water (72 mN/m at 25°C) decreases to 63 mN/m in a solution of SDS of concentration 1.7 mmol/L. The large decrease in surface tension suggests that SDS molecules are concentrated at the surface, as otherwise there should be very little change in surface tension. This means that the concentration of SDS at the surface is much higher than in the bulk. The molar ratio of SDS:water in the bulk is 0.002:55.5. At the surface, the ratio will be expected to be of a completely different value, as found from the value of Γ (ratio is 1000:1). This is also obvious when considering that foam bubbles form on solutions with very low surface-active agent concentrations. The foam bubble consists of a bilayer of surface-active agent with water inside. In fact, it is easy to consider the state of surfactant solutions in terms of molecular ratios (Chapter 7).

For an ionizing surfactant (detergent), the form of Gibbs equation is less certain. If we use SDS ($C_{12}H_{25}SO_4Na = C_{12}H_{25}SO_4^- + Na^+$), for example, since it is a strong electrolyte, it can be considered to dissociate completely:

$$C_{12}H_{25}SO_4Na = C_{12}H_{25}SO_4^- + Na^+$$

$$SDS = DS^- + S^+ \tag{3.32}$$

The appropriate form of the Gibbs equation will be

$$-d\gamma = \Gamma_{DS-}\,d\mu_{DS-} + \Gamma_{S+}d\mu_{S+} \tag{3.33}$$

where surface excess, Γ_i, represents for each species in the solution, for example: DS^- and S^+ are included. This equation relates the observed change in surface tension to the changes in the chemical potential of the respective solutes (here: DS^-; S^+). If one expands the chemical potential terms, then one obtains

$$-d\gamma = RT\left[\Gamma_{DS} - d\left(\ln C_{DS-}\right) + \Gamma_S + d\left(\ln C_{S+}\right)\right] \tag{3.34}$$

$$= RT\left(\Gamma_{DS} - dC_{DS-}/C_{DS-} + \Gamma_S + dCs + /C_{S+}\right) \tag{3.34a}$$

Assuming electrical neutrality is maintained in the interface, then one gets

$$\Gamma_{SDS} = \Gamma_{DS-} = \Gamma_{S+} \tag{3.35}$$

and

$$C_{SDS} = C_{DS-} = C_{S+} \tag{3.36}$$

which, on substitution in Equation 3.3, gives

$$-d\gamma = 2\ RT\Gamma_{SDS}d\left(\ln C_{SDS}\right) \tag{3.37}$$

$$= \left(2RT/C_{SDS}\right)\Gamma_{SDS}\ dC_{SDS} \tag{3.38}$$

and one obtains

$$\Gamma_{SDS} = -1/2\left(R\ T\right)\ d\gamma/\left(d\ln C_{SDS}\right) \tag{3.39}$$

In the case where the ion strength is kept constant, that is, in the presence of added NaCl, then the equation becomes

$$\Gamma_{SDS} = -1/\left(RT\right)\ d\gamma/\left(d\ln C_{SDS}\right) \tag{3.40}$$

Comparing Equation 3.38 with Equation 3.40, it will be seen that they differ by a factor of 2, and that the appropriate form will need to be used in experimental test of the Gibbs equation. It is also quite clear that any partial ionization would need necessary revisions to the general Gibbs equation. The adsorption of detergents at the surface of a solution can be estimated by Gibbs equation (Chattoraj and Birdi, 1984; Adamson and Gast, 1997; Somasundaran, 2015; Birdi, 2016). It is convenient to plot γ versus the log ($C_{detergent}$). From γ versus $C_{alkyl\ sulfate}$ data, one finds the following data as obtained from the Gibbs equation:

	Concentration (mol/L)	$\Gamma_{Salkylsulfate}$	A (area/molecule) 10^{-12} mol/cm²
NaC$_{10}$sulfate	0.03	3.3	50 Å²
NaC$_{12}$sulfate	0.008	3.4	50 Å²
NaC$_{14}$sulfate	0.002	3.3	50 Å²

From the data plots of γ versus concentration (where the slope is related to the surface excess, $\Gamma_{Salkylsulfate}$), one can estimate the value of the area/molecule of the adsorbed SAS. The area/molecule values indicate that the molecules are aligned *vertically* on the surface, irrespective of the alkyl chain length. If the molecules were oriented flat, then the value of the area/mole would be much larger (approximately 100 Å^2). The fact that the alkyl chain length has no effect on the area also proves this assumption. These conclusions have been verified from spread monolayer studies. Furthermore, one also finds that the polar group, that is, $-SO^{4-}$, would occupy something like 50 Å^2. In Chapter 4, it will be shown that other studies confirm that the area per molecule is approximately 50 Å^2. This is the only indirect method by which one can determine the surface structure.

Gibbs adsorption equation basically relates the chemical potential of a solvent and a solute (or multisolutes). The solute is present either as *excess* (if there is an excess surface concentration) if the solute decreases the γ, or as a *deficient* solute concentration (if surface tension is increased by the addition of the solute). It can be further explained that if we consider water as a system, to which a surfactant, such as SDS, is added, the molecules at the surface will change as follows:

PURE WATER: bulk water molecule = w; surface water molecule = **w**:

- *wwwwwwwwwwwwwwwwwwwwwwwwww*
- wwwwwwwwwwwwwwwwwwwwwwwwww
- wwwwwwwwwwwwwwwwwwwwwwwwww

WATER PLUS SDS (bulk SDS = S; surface SDS = **s**. (SDS = 2 g/L):

- *ssswsswsssswsswssswssswssswsswsssswsswsss*
- swwwswwwswwwsww w wsw w wsw w wsww wws
- wwwwwwswwwwwwwswwwwwwwswwww

This shows that surface tension of pure water of 72 mN/m decreases to 30 mN/m by the addition of 2 g/L of SDS. Thus, the surface of SDS solution is mostly a monolayer of SDS plus some bound water. The ratio of water:SDS in the system is roughly as follows:

- In bulk phase: 55 mole water:8 mmole SDS
- At the surface: roughly 100 mole SDS:1 mole water

This description is in accord with the decrease in γ of the system (and analyses by other methods: Adamson and Gast, 1997; Chattoraj and Birdi, 1984). Investigations have shown that if one carefully sucked a small amount of surface solution of a surfactant, then one can estimate the magnitude of Γ. The concentration of SAS was found to be 8 μmol per milliliter. The concentration in the bulk phase was 4 μmol/L. The data show that the surface excess is 8 μmol/mL − 4 μmol/mL = 4 μmol/mL. Furthermore, this indicates that when there is 8 μmol/L in the bulk of the solution, then at the surface the SDS molecules completely cover the surface. The consequence of this is that at a higher concentration than 8 μmol/L, no more adsorption at the interface of SDS takes place. Thus, γ *remains constant* (almost). This means that the surface is completely covered with SDS molecules. The area per molecule data (as found to be 50 Å^2) indicates that the SDS molecules are oriented with the SO^{4-} groups pointing toward the water phase while the alkyl chains are oriented away from the water phase.

This means that if one collected the foam continuously, then more and more SAS will be removed. This method of bubble–foam separation has been used to purify wastewater of SAS. It is especially useful when there are very minute amounts of SAS (e.g., dyes in the printing industry or pollutants in wastewater). It is economical and free of any chemicals or filters. In fact, if the pollutant is very expensive or poisonous, then this method can have many advantages over other methods. This property is unique for all surfactants (Birdi, 2014, 2016; Somasundaran, 2015).

Example

It is useful to estimate the amount of SDS in a bubble of radius 1 cm. Assuming that there is an almost negligible amount of water in the bilayer of the bubble, the surface area of the bubble can be used to estimate the amount of SDS. The known data are as follows:

Radius of bubble = 1 cm
Surface area = (4 Π 1^2) 2 = 25 cm^2 = 25 × 10^{16} Å^2
Area per SDS molecule (as measured from other methods) = 55 Å^2
Number of SDS molecules per bubble = 0.5 × 10^{16} molecules
Amount of SDS per bubble = 0.5 × 10^{16}/6 × 10^{23} g
= 0.01 μg SDS

It can thus be seen that it would require *100 million* bubbles to remove 1 g of SDS from the solution, or 100,000 bubbles to remove 1 mg of SDS (in a solution of 1 mg/L = 1 ppm). Pollutants are generally in this range of concentration (ppm). Since bubbles can be easily produced at very fast rates (ca. 100–1000 bubbles per minute), this is not a big hindrance. Consequently, any kind of other SAS (such as pollutants in industry) can thus be removed by foaming (Birdi, 2016).

The latter example shows some useful applications of bubble formation in the removal of any SAS in water. For example, if SDS is added to a solution of water with any organic molecule in water, then the concentration of the former in the bubble will be higher in the bubble than in the bulk. Hence, one can remove the organic molecules by collecting bubbles (Chapter 7). Different experiments have been carried out to verify the Gibbs adsorption theory (Adamson and Gast, 1997; Chattoraj and Birdi, 1984; Birdi, 2016). One of these methods was carried out by removing with a microtone blade the thin surface layer of a surfactant solution. This is almost the same as the process of bubble extraction carried out by the careful suction of the surface layer of solution. The surface excess data for a solution of SDS were found to be acceptable. The experimental data was 1.57 10 $^{-18}$ mole cm^{-2}, while, from the Gibbs adsorption equation, one expected it to be 1.44 × 10 $^{-18}$ mole cm^{-2}. It is important to mention that no other direct method to measure these quantities exists.

Example: Aqueous solution of CTAB shows the following data (25°C):

$$\gamma = 47 \text{ mN/m}, C_{ctab} = 0.6 \text{ mmol/L}$$

$$\gamma = 39 \text{ mN/m}, C_{ctab} = 0.96 \text{ mmol/L}$$

By using Equation 3.40, one gets

$$d \gamma/d \log(C_{ctab}) = (47-39)/(\log(0.6) - \log(0.96))$$

$$=8/(-0.47) = 17$$

In the following data, the rate of change of surface tension of water is shown for various additives. The most interesting is that the inorganic solutes give an increase in surface tension (excepting HCl).

Change in surface tension of water on addition of various substances (mN/mole) (25°C):

SOLUTE	$d\gamma/dC_{solute}$
HCl	−0.3
LiCl	+1.8
NaCl	+1.8
CsCl	+1.54
CH_3COOH	−38

It is thus seen that the interfacial properties of these different solutes will be dependent on the surface concentration of the solute. Extensive analyses of these aspects have been given in the literature (Chattoraj and Birdi, 1984; Adamson and Gast, 1997; Birdi, 2016; Somasundaran, 2015).

3.4.1 KINETIC ASPECTS OF SURFACE TENSION OF DETERGENT AQUEOUS SOLUTIONS

Without exception, one needs kinetic information in the analysis of any phenomenon. In the present case, one would like to ask how fast the surface tension of a detergent solution reaches equilibrium. If one pours a detergent solution into a container, then the instantaneous concentration of the detergent will be uniform throughout the system, that is, it will be the same in the bulk and at the surface. Since (in general) the concentration of the SAS is very low, then the surface tension of the solution will be the same as of pure water (i.e., 72 mN/m at 25°C). This is because the magnitude of the surface excess, *at time zero*, is zero (i.e., $\Gamma_{time=0} = 0$).

However, it is found that the freshly formed surface of a detergent solution exhibits varying rates of change in surface tension with time. A solution is uniform in solute concentration in the bulk phase, until a surface is created. At the *surface*, the SAS will accumulate and the surface tension will decrease with time. In some cases, the rate of adsorption at the surface is very fast (less than a second) while in other cases, it may take longer. One finds that the freshly created aqueous solution shows

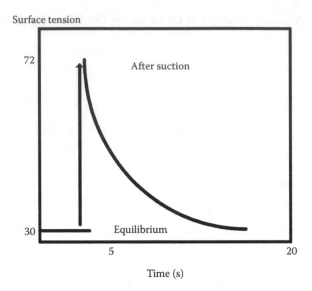

FIGURE 3.13   Kinetics of surface tension: change (of a typical surfactant aqueous solution) in equilibrium surface tension after suction at the surface (see text for details).

the surface tension of almost pure water, that is, 70 mN/m. However, surface tension starts to decrease rapidly and reaches an equilibrium value (which may be lower than 30 mN/m) after a given time (a few seconds). If, after this equilibrium stage, a slight suction at the surface is applied, then the magnitude of the surface tension increases to that of pure water (ca. 70 mN/m) (Figure 3.13). Thereafter, the surface tension of the solution drops rapidly to its equilibrium value. Experiments show that this process can be repeated several times. In general, this phenomenon has no consequences. However, in some cases where a fast cleaning process is involved, one must consider the kinetic aspects. In fact, the formation of foam bubbles as one pours the solution is indicative that surface adsorption is indeed very fast (as pure water does not foam on shaking).

Especially, in the case of high molecular weight SAS (such as proteins), the period of change in surface tension is found to be sufficiently prolonged to allow easy observations. This arises from the fact that proteins are surface active. All *proteins* behave as SAS because of the presence of hydrophilic–lipophilic properties (imparted from the different *polar* [such as glutamine and lysine] and *apolar* [such as alanine, valine, phenylalanine, and iso-valine] amino acids). Proteins have been extensively investigated as regards their polar–apolar characteristics as determined from surface activity (Chattoraj and Birdi, 1984; Birdi, 2016; Tanford, 1980). Based on simple diffusion assumptions, the rate of adsorption at the surface, Γ, can be expressed as

$$d\Gamma/dt = (\mathbf{D}/\pi)^2 C_{bulk} t^{-2} \tag{3.41}$$

which, on integration, gives

$$\Gamma = 2\, C_{bulk} \left(\mathbf{D}t/\pi\right)^2 \tag{3.42}$$

where:

D   is the diffusion constant coefficient

C_{bulk} is the bulk concentration of the solute

The procedure used is to apply suction at the surface and the fresh surface is created instantaneously (Birdi, 1989). The magnitude of γ increases to pure water (72 mN/m) and decreases with time as Γ increases (from initial value of zero) (Figure 3.13). This experiment actually verifies the various assumptions as made in the Gibbs adsorption equation. Experimental data show good correlation to this equation when t is very small.

3.5 SOLUBILIZATION (OF ORGANIC WATER-INSOLUBLE MOLECULES) IN MICELLES

In many everyday needs, one has to be able to apply organic water-insoluble compounds in industry and biology. Or, in some cases, such as pollution, one needs to know the solubility characteristics of pollutants in water. It has been found that micelles (both ionic and nonionic) behave as a micro-phase, where the *inner core* behaves as (liquid) alkane, while the surface area behaves as a polar phase (Tanford, 1980; Chattoraj and Birdi, 1984; Birdi, 1997, 2014, 2016). The inner core is also found to exhibit liquid-alkane-like characteristic. The inner core thus has been found to exhibit alkane-like properties while being surrounded by a water phase. In fact, micelles are considered nanostructures (with radii of molecular dimensions) (Tanford, 1980; Birdi, 1997, 2016; Somasundaran, 2015). What this then suggests is that one can design surfactant solution systems in water, which can have both aqueous and alkane-like properties. This unique property is one of the main applications of surfactant micelle solutions in all kinds of systems. Furthermore, in ionic surfactant micelles, one can also create *nano-reactor* systems. In nano-reactors, the counter-ions are designed to bring two reactants into very close proximity (due to EDL). These reactions would otherwise have been impossible (Scheludko, 1966; Birdi, 2007, 2016). The most useful characteristic of a micelle arises from its inner (alkyl chains) part (Figure 3.14).

The micelles thus provide a unique method of a nano-sized phase in aqueous solutions. The inner part consists of alkyl groups, which are closely packed. It is known that these clusters behave as *liquid paraffin* ($C_n H_{2n+2}$). The alkyl chains are thus not fully extended. Hence, one would expect that this inner hydrophobic part of a micelle should exhibit properties, which are common for (liquid) alkanes, such as the ability to solubilize all kinds of water-insoluble organic compounds. This characteristic has been verified by experimental data. The solute enters the alkyl core of the micelle, and it swells. The chemical equilibrium is reached when the ratio between moles solute:moles detergent is reached corresponding to the thermodynamic value.

Size analyses of micelles (by using light-scatter) of SDS have indeed shown that the radius of the micelle is almost the same as the length of the SDS molecule. However, if the solute interferes with the outer polar part of a micelle, then the micelle system may change, such that the CMC and other properties change. This is observed in the case of dodecanol addition to SDS solutions. However, very small additions of solutes show very little effect on CMC. A typical system is analyzed

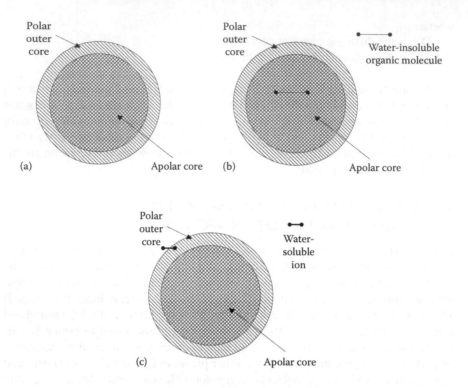

FIGURE 3.14 Micelle structure: (a) inner part = liquid paraffin-like; outer polar part; (b) solubilization of apolar molecule; (c) binding of counter-ion to the polar part.

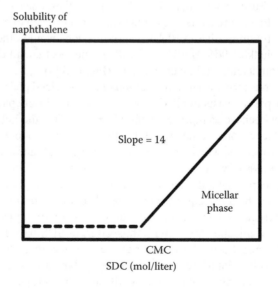

FIGURE 3.15 Solubilization of naphthalene in SDS solutions (a typical micelle solubilization example) (25°C).

here to delineate the solubility mechanism. The data in Figure 3.15 show the change in solubility of naphthalene in SDS aqueous solutions.

Below CMC, the amount of naphthalene dissolved remains constant, which corresponds to its solubility in pure water. Below CMC, the monomers have no effect on the solubility. The slope of the plot above CMC corresponds to 14 mole SDS:1 mole naphthalene. This ratio is unique for each solute, and is not dependent on the CMC of the system (i.e., the addition of electrolytes, etc.). It is seen that at the CMC, the solubility of naphthalene abruptly increases. This is because all micelles can solubilize water-insoluble organic compounds. A more useful analysis can be carried out by considering the thermodynamics of this solubilization process. At equilibrium, the chemical potential of a solute (naphthalene, etc.) will be given as

$$\mu_s^s = \mu_s^{aq} = \mu_s^M \tag{3.43}$$

where:

μ_s^s is the chemical potential of the solute in the solid state
μ_s^{aq} is the chemical potential of the solute in the aqueous phase
μ_s^M is the chemical potential of the solute in the micellar phase

It must be noted that in these micellar solutions, we will describe the system in terms of aqueous phase and micellar phase. The standard free energy change involved in the solubilization, ΔG°_{so}, is given as follows:

$$\Delta G^\circ_{so} = - RT \ln\left(C_{s,M} / C_{s,aq}\right) \tag{3.44}$$

where:

$C_{s,aq}$ is the concentration of the solute in the aqueous phase
$C_{s,M}$ is the concentration of the solute in the micellar phase

The free energy change is the difference between the energy when a solute is transferred from solid (or liquid) state to the micelle interior. It has been found from many systematic studies that ΔG°_{so} is dependent on the chain length of the alkyl group of the surfactant. The magnitude of ΔG°_{so} changes by −837 J (−200 cal)/mole, with the addition of -CH$_2$- group. In most cases, the addition of electrolytes to the solution has no effect (Birdi, 1982, 1999, 2003, 2016). The kinetics of solubilization has an effect on its applications (Birdi, 2003). Another important aspect is that the slope in Figure 3.18 corresponds to $(1/C_{sM})$. This allows one to determine moles of SDS required to solubilize one mole of solute. Different analyses of various solutes (water-insoluble organic molecules) in SDS aqueous micellar systems showed that

Solute	Ratio (SDS:solute) in the Micelle Phase
Naphthalene	14 moles SDS/mole naphthalene
Anthracene	780 moles SDS/mole anthracene
Phenathrene	47 moles SDS/mole phenathrene

It is also found that the solubility data of these water-insoluble organic compounds are related only to the chain length of the different surfactants. One can conclude that this shows that the solubilization takes place in the interior of the micelle, and that the interior is liquid-alkane-like and similar in all micelles. The rate of solubilization has been investigated (Birdi, 2003, 2016). These data thus allow one to determine (quantitatively) the range of solubilization in any such application. Industrial applications such as pharmaceuticals, agricultural sprays, paints, etc., require such information on solubilizing water-insoluble organic compounds. The dosage of any substance is based on the amount of material per volume of a solution. This thus also shows that wherever detergents are employed, their major role (besides achieving lower surface tension) is the solubilization of any water-insoluble organic compound. This process assists in the cleaning or washing function. In some cases, such as bile salts, the solubilization of lipids (especially lecithins) gives rise to some complicated micellar structures. Due to the formation of mixed lipid–bile salt micelles, one observes changes in the CMC and aggregation number. This has major implications for the use of bile salts in biology.

4 Surface Chemistry of Solid Surfaces
Adsorption–Desorption Characteristics

4.1 INTRODUCTION

Solid surfaces exhibit some specific characteristics that are much different to liquid surfaces (Adamson and Gast, 1997; Scheludko, 1966; Chattoraj and Birdi, 1984; Birdi, 2016; Somasundaran, 2015). In most processes where solids are involved, the primary step is dependent on the surface property of the solid. One finds a large variety of applications where the surface of a solid plays an important role, for example in catalysis, on road surfaces, in oil and gas reservoirs (shale reservoirs), and on substances such as active charcoal, talcum powder, cement, sand, plastics, wood, glass, clothes, hair, skin, etc. Solids are rigid structures that resist any stress effects. In all porous solids, the flow of gas or liquid oil means that interfaces (liquid–solid), are involved. Hence, in such a system, the molecular interactions in the various phases of a surface's chemistry are of primary importance. The latter is especially the case in complex systems, such as shale oil/gas reservoirs. In these systems, there are some specific stages where the surface characteristics (of solids and liquids) will be of importance. Many important technical processes and natural ones such as earthquakes, in everyday life are dependent on the characteristics of rocks, etc., in the interior of the earth. To understand these processes, we must understand the surface forces on solid interfaces (solid–gas, solid–liquid, $solid_1$–$solid_2$).

Currently, a major example of this subject is the shale oil and gas reservoir systems. At a microscopic level, one finds that these reservoirs are dependent on both the shale structure and the gas type. Some of these reservoirs are source-reservoir systems. The gas in shale is found to have been generated from organic material (plants, etc.) (Appendix I) (Calvin, 1969). The surface of shale will thus be expected to be the most important factor in recovery phenomena. Shale consists of both inorganic and varying amounts of organic substances (Calvin, 1969) (Appendix I).

A general surface chemistry of solid surfaces is delineated here. The surface chemistry of solids can be described based on the classical theories (Adamson and Gast, 1997; Birdi, 2002, 2016; Somasundaran, 2015). Another example is the corrosion of metals, which initiates at surfaces, thus requiring treatments that are based on surface properties (Perenchio, 1994; Adamson and Gast, 1997; Roberge, 1999; Ahmad, 2006; McCafferty, 2010; Zinola, 2010). As described in the case of liquid

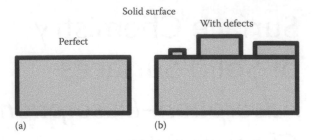

FIGURE 4.1   Solid surface molecules defects: (a) perfect crystal; (b) surface with defects.

surfaces, analogous analyses of solid surfaces can be carried out. The molecules at the solid surfaces are not under the same force field as in the bulk phase (Figure 4.1).

Experiments show that the solid surface is the most important characteristic. The differences between perfect surfaces and surfaces with *defects* are very obvious in many everyday observations (Figure 4.1). For example, the shine of all solid surfaces increases as the surface becomes smoother. Furthermore, the friction decreases between two solid surfaces as the solid surfaces become smoother. Solid materials were the first to be analyzed at the molecular scale (using x-ray diffraction, etc.). This led to the understanding of the structures of solid substances and the crystal atomic structure. This is because while molecular structures of solids can be investigated by such methods as x-ray diffraction, the same analyses of liquids are not that straightforward (Gitis and Sivamani, 2014). These analyses have also shown that surface defects exist at the molecular level. As pointed out in the case liquids in the previous chapter, one must also consider that when the surface area of a solid powder is increased by grinding (or some other means), then *surface energy* (energy supplied to the system) is needed. Of course, due to the energy differences between solid and liquid phases, these processes will be many orders of magnitude different from each other. A substance in a liquid state retains some structure, which is similar to its solid state, but in a liquid state the molecules exchange places. The average distance between molecules in the liquid state is roughly 10% larger than in their solid state (see Chapter 1). It is thus desirable at this stage to consider some of the basic properties of liquid–solid interfaces. The surface tension of a liquid becomes important when it comes in contact with a solid surface. The interfacial forces, which are present between a liquid and a solid, can be estimated by studying the shape of a drop of liquid placed on any smooth solid surface (Figure 4.2). The shape (or the angle) of the drop of liquid on different solid surfaces is found to be different. The balance of forces as indicated, has been extensively analyzed, which relates different forces at the solid–liquid boundary, and the contact angle, θ, as follows (Young's equation) (Adamson and Gast, 1997; Chattoraj and Birdi, 1984, 1997a, 2002, 2016):

$$(\text{Surface tension of solid})$$

$$= \text{Surface tension of solid / liquid} + \text{Surface tension of liquid } (Cos(\theta)) \quad (4.1)$$

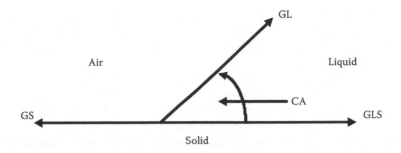

FIGURE 4.2 The state of equilibrium between the surface tensions of liquid (GL), solid (GS), and liquid/solid (GLS) (at a contact angle [CA]).

$$\gamma_S = \gamma_L Cos(\theta) + \gamma_{SL} \tag{4.2}$$

$$\gamma_L Cos(\theta) = \gamma_S - \gamma_{SL} \tag{4.3}$$

where the different surface forces (surface tensions) are γ_S for solid, γ_L for liquid, and γ_{SL} for solid–liquid interface.

This is found to be the most fundamental relation between a solid surface and its interaction with fluid. The relation of Young's Equation 4.2 is essentially based on physics laws of force equilibrium. At equilibrium, the magnitude of the contact angle relates to the three surface forces (Figure 4.2).

In Equation 4.3 only the geometrical force balance is considered in the X–Y plane. It is assumed that the liquid does not affect the solid surface (in any physical sense). This assumption is safe in most cases. However, in very special cases, if the solid surface is soft (such as a contact lens), then one will expect that tangential forces will also need to be included in this equation (Birdi, 2016).

4.2 WETTING PROPERTIES OF SOLID SURFACES

The degree of wetting when a liquid comes in contact with a solid surface is the most common phenomenon in everyday life (washing and detergents, water flow underground, hydraulic fracking process, rainwater seepage and floods, cleaning systems, water flow in rocks, printing technology, etc.). The liquid and solid surface interface can be described by considering a classical example. The wetting of solid surfaces is well known when considering the difference between Teflon and metal surfaces. To understand the degree of *wetting* between the liquid, L, and the solid, S, it is convenient to rewrite Equation 4.3 as follows:

$$Cos(\theta) = (\gamma_S - \gamma_{LS}) / \gamma_L \tag{4.4}$$

which would then allow one to understand the variation of γ with the change in the other terms. The latter is important because complete wetting occurs when there is no finite contact angle, and thus $\gamma_L <> \gamma_S - \gamma_{LS}$. However, when $\gamma_L > \gamma_S - \gamma_{LS}$, then

$Cos(\theta) < 1$, and a finite contact angle is present. The latter is the case when water, for instance, is placed on a hydrophobic solid, such as Teflon, polyethylene (PE), or paraffin. The addition of surfactants to water, of course, reduces γ_L; therefore, θ will decrease on the introduction of such surface-active substances (Adamson and Gast, 1997; Chattoraj and Birdi, 1984; Birdi, 1997, 2002, 2016; Zhang et al., 2012). The state of a fluid drop under dynamic conditions, such as evaporation becomes more complicated (Birdi et al., 1989; Birdi and Vu, 1989). However, in this text, one is primarily interested in the spreading behavior when a drop of one liquid is placed on the surface of another liquid, especially when the two liquids are immiscible. Understanding this phenomenon is important in understanding the case of oil spills in oceans, and so on.

The spreading phenomenon was analyzed by introducing a quantity, spreading coefficient, $S_{a/b}$, defined as (Harkins, 1952; Adamson and Gast, 1997; Birdi, 2002, 2016)

$$S_{a/b} = \gamma_a - (\gamma_b + \gamma_{ab})   \qquad (4.5)$$

where:

$S_{a/b}$ is the spreading coefficient for liquid **b** on liquid **a**

γ_a and γ_b are the respective surface tensions

γ_{ab} is the interfacial tension between the two liquids

If the value of $S_{b/a}$ is positive, spreading will take place spontaneously, while if it is negative, liquid **b** will rest as a lens on liquid **a**.

However, the value of γ_{ab} needs to be considered as the equilibrium value, and therefore if one considers the system at nonequilibrium, then the spreading coefficients will be different. For instance, the instantaneous spreading of benzene is observed to give a value of $S_{a/b}$ as 8.9 dyne/cm; therefore, benzene spreads on water. On the other hand, as the water becomes saturated with time, the value of (water) decreases and benzene drops tend to form lenses. Short-chain hydrocarbons such as n-hexane and n-hexene also have positive initial spreading coefficients and spread to give thicker films. Longer-chain alkanes on the other hand, do not spread on water, for example, the $S_{b/a}$ for $C_{16}H_{34}$ (hexadecane)/water is ca. 1.3 dyne/cm at 25°C. It is also obvious that since impurities can have very drastic effects on the interfacial tensions in Equation 4.5, the value of $S_{a/b}$ would be expected to vary accordingly (Table 4.1).

The spreading of a solid (polar organic) substance, for example, cetyl alcohol ($C_{18}H_{38}OH$), on the surface of water has been investigated in some detail (Gaines, 1966; Adamson and Gast, 1997; Birdi, 2003). Generally, however, the detachment of molecules of the amphiphile into the surface film occurs only at the periphery of the crystal in contact with the air–water surface. In this system, the diffusion of the amphiphile through the bulk water phase is expected to be negligible, because the energy barrier now includes not only the formation of a hole in the solid, but also the immersion of the hydrocarbon chain in the water. It is also obvious that diffusion through the bulk liquid is a rather slow process. Furthermore, the value of $S_{a/b}$ would be very sensitive to such impurities as regards the spreading of one liquid on another.

Another example is the addition of surfactants (detergents) to a fluid. This dramatically affects its wetting and spreading properties. Thus, many technologies

---

TABLE 4.1
Calculation of Spreading Coefficients, $S_{a/b}$, for Air–Water Interfaces (20°C)

Oil	$\gamma_{w/a} - \gamma_{o/a} - \gamma_{o/w} = S_{a/b}$	Conclusion
n-$C_{16}H_{34}$	$72.8 - 30.0 - 52.1 = -0.3$	Oil will not spread
n-Octane	$72.8 - 21.8 - 50,8 = +0.2$	Oil will just spread
n-Octanol	$72.8 - 27.5 - 8.5 = +36.8$	Oil will spread

Note: a = air; w = water; o = oil.

---

utilize surfactants for control of wetting properties (Birdi, 1997). The ability of surfactant molecules to control wetting arises from their *self-assembly* at the liquid–vapor, liquid–liquid, solid–liquid, and solid–air interfaces and the resulting changes in the interfacial energies. These interfacial self-assemblies exhibit rich structural detail and variation. The molecular structure of the self-assemblies and the effects of these structures on wetting or other phenomena remain topics of extensive scientific and technological interest.

As an example, in the case of oil spills on the sea, these considerations become very important. The treatment of such pollutant systems requires knowledge of the state of the oil. The thickness of the oil layer will be dependent on the spreading characteristics. The effect on ecology (birds, plants, etc.) will depend on the spreading characteristics. Young's equation at liquid$_1$–solid–liquid$_2$ has been investigated for various systems. It is found in such systems where the liquid$_1$_solid–liquid$_2$ surface tensions meet at a given contact angle. For example, the contact angle of a water drop on Teflon is 50^u in octane (Chattoraj and Birdi, 1984) (Figure 4.3):

water (liquid) Teflon (solid) octane (liquid)

In this system, the contact angle, θ, is related to the different surface tensions as follows:

$$\gamma_{s\text{-octane}} = \gamma_{water\text{-}s} + \gamma_{octane\text{-}water}\cos(\theta) \tag{4.6}$$

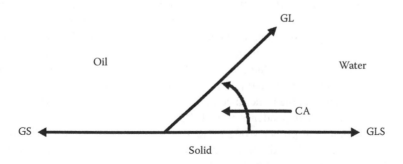

FIGURE 4.3 Contact angle (CA) at water–Teflon–octane interface.

or

$$Cos(\theta) = (\gamma_{\text{s-octane}} - \gamma_{\text{water-s}}) / \gamma_{\text{octane-water}} \qquad (4.7)$$

This gives the value of $\theta = 50°$ when using the measured values of $\gamma_{\text{s-octane}} - \gamma_{\text{water-s}} / \gamma_{\text{octane-water}}$. The experimental value of $\theta(=50°)$ is the same. This analysis shows that the assumptions made in the derivation of Young's equation are valid. This example is analogous to the oil–shale–brine phase. Current investigations on the wetting behavior of shale rocks have been reported (Borysenko et al., 2008).

4.2.1 HYDRAULIC FRACTURE FLUID INJECTION AND WETTABILITY OF SHALES

In the shale reservoirs, the water phase (hydro-fracking fluid) is injected to create fractures (under high pressure) (Borysenko et al., 2008; Shabro et al., 2012). This process indicates that the *wetting properties* of the system will be among the main parameters of interest. The hydrophilic or hydrophobic (hydrophilic–hydrophobic balance [HLB]) characteristics of the reservoir have been investigated (Borysenko et al., 2008). The reservoir was classified as water-wet, oil-wet, or mixed-wet shales. Contact angle measurements were made to determine these characteristics. Different rocks were investigated (quartz, kaolinite, montmorillonite, shale). The differences in contact angle data were related to the water-wet or oil-wet properties of the solid.

4.2.1.1 Hydraulic Fracturing Fluid (Water Phase) and Reservoir

The fluid (water plus additives) is injected into the reservoir. In order to keep the fractures open after the fracturing process, silica particles (or similar) are used. Pressure is released after the fracturing step. The injected fluid (mixed with some reservoir saline water) flows back up through the well bore. However, most of the injected fluid (over 60%) remains in the reservoir due to adsorption and so on. The fluid that is recovered is referred to as *flow-back*.

Furthermore, the degree of fluid recovery (after the fracturing step) will be dependent on the wetting properties of the shale. Studies have been carried out to investigate the wetting characteristics and fracture process (Borysenko et al., 2008; Aderibigbe, 2012; Donaldson et al., 2013). After the fracking process step, the gas (desorbed) pushes the water phase toward the borehole (i.e., after the pressure decreases). The mechanism of this phenomenon has been found to be related to the wetting characteristics of the shale. This phenomenon has an important relation to the amount of water recovered and other related aspects. Furthermore, the total organic content (TOC) relates to the wetting characteristics. TOC is organic material with less wetting characteristics than the mineral content (high wetting) (Appendix I). The degree of wetting will thus be dependent on the TOC.

The most important property of a surface (solid or liquid) is its interaction with other materials (gases, liquids, or solids). All interactions in nature are governed by different kinds of molecular forces (such as Van der Waals, electrostatic, hydrogen bonds, dipole–dipole interactions). Based on various molecular models, the surface tension, γ_{12}, between two phases with γ_1 and γ_2, was given as (Adamson and Gast, 1997; Ross, 1971; Chattoraj and Birdi, 1984; van Oss, 2006; Birdi, 2016; Hansen, 2007):

$$\gamma_{12} = \gamma_1 + \gamma_2 - 2\Phi_{12}(\gamma_1\gamma_2)^2 \tag{4.8}$$

where Φ_{12} is related to the interaction forces across the interface. The latter parameter is easy to expect, since the interface forces will depend on the molecular structures of the two phases. In the case of systems such as alkane (or paraffin)–water, their Φ_{12} is found to be equal to unity in Equation 4.8. Φ_{12} is unity since an alkane molecule exhibits no hydrogen-bonding property, while water molecules are strongly hydrogen bonded. It is thus found that in all liquid–solid interfaces, there will be present different apolar (dispersion) forces + polar (hydrogen-bonding; electrostatic forces). Hence, all liquids and solids will exhibit γ of different kinds:

- Liquid surface tension:

$$\gamma_L = \gamma_{L,D} + \gamma_{L,P} \tag{4.9}$$

- Solid surface tension:

$$\gamma_S = \gamma_{S,D} + \gamma_{S,P} \tag{4.10}$$

This means that γ_S for Teflon arises only from *dispersion* (SD) forces. On the other hand, a glass (or quartz, rock, silica, or other mineral) surface shows γ_S, which will be composed of both γ_{SD} and γ_{SP}. Hence, the main difference between a Teflon and a glass surface will arise from the γ_{SP} component of glass. This has been found to be of importance in the case of the application of adhesives. The adhesive used for glass will need to bind to a solid with both polar and apolar forces. The values of γ_{SD} for different solids as determined from these analyses are given next.

Solid surface tensions:

Solid	γ	γ_{SD}	γ_{SP}
Teflon	19	19	0
Polypropylene	28	28	0
Polycarbonate	34	28	6
Nylon 6	41	35	6
Polystyrene	35	34	1
PVC	41	39	2
Kevlar49	39	25	14
Graphite	44	43	1

The estimation of the different solid forces has been found to correlate with various practical systems (such as adhesion, glue, friction, etc.).

The solid surface tension related to polar and nonpolar forces is of particular significance. Furthermore, the *asymmetrical* forces acting at the surfaces of liquids are much shorter than those expected on solid surfaces. This is due to the high energies that stabilize the solid structures.

4.3 SURFACE TENSION (γ_{SOLID}) OF SOLIDS

As described in Chapter 2, the molecules at the surface of a liquid are under tension due to asymmetrical forces. However, in the case of solid surfaces, one may not envision this kind of asymmetry as clearly. Nevertheless, simple observation can help one to realize that such surface tension asymmetry also exists in solids. For instance, let us analyze the state of a drop of water (ca. 10 μL) as placed on two different smooth solid surfaces, for example, Teflon (nonwetting: $\theta > 90°$) and glass (wetting: $\theta < 90°$). One finds that the contact angles are different (Figure 4.4). The contact angle is defined as indicated in Figure 4.4. Since the surface tension of water is the same in the two systems, then the difference in contact angles can only arise due to the surface tension of solids being different.

The surface tension of liquids can be measured directly (as described in Chapter 2). However, it is known that this is not possible in the case of solid surfaces. The surface tension of solids is thus estimated through indirect methods (such as data of liquid–solid contact angle). Experiments show that when a liquid drop is placed on a solid surface, the contact angle, θ, indicates that the molecules interact across the interface. This thus indicates that these data can be used to estimate the surface tension of solids. The fracking process is also related to the surface tension of shale and the surface tension of the fracking aqueous solution.

4.4 CONTACT ANGLE (θ) OF LIQUIDS ON SOLID SURFACES

As already mentioned, a solid in contact with a liquid leads to interactions related to the different surface forces (i.e., surface tensions of liquid and solid) involved. The solid surface is being brought in contact with the surface forces of the liquid (surface tension of liquid). This phenomenon can be conveniently investigated as follows: If a small drop (a few microliters) of water is placed on a smooth surface of Teflon or glass (Figure 4.4), one finds that the shapes of these drops are different. The reason being that there are three surface forces (tensions), which, at the thermodynamic equilibrium, give rise to a contact angle, θ. Young's equation describes the interplay of forces (liquid surface tension; solid surface tension; liquid–solid surface tension) at the three-phase boundary line. It is regarded as if these forces interact along a line. Experimental data show that this is indeed true. The magnitude of θ is thus only

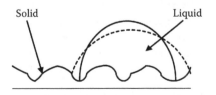

FIGURE 4.4   Apparent or true contact angle of a liquid drop on a rough solid surface.

dependent on the molecules nearest the interface, and is independent of molecules much further away from the contact line. Furthermore, one defines that

- When θ is less than 90°, the surface is wetting (such as water on glass).
- When θ is greater than 90°, then the surface is nonwetting (such as water on Teflon).
- When θ is equal to 0°, this is observed when there is strong attraction between the liquid and solid molecules.

Most important to note here is that by treating a glass surface with suitable chemicals, the surface can be rendered hydrophobic. Also, with suitable treatments, the nonwetting (hydrophobic) surface of polystyrene (PS) can be made more wetting (hydrophilic) (Birdi, 1989). It was found that by treating PS with sulfuric acid, the surface becomes more hydrophilic (and can attract binding of bacteria, etc.). This is the same technology as used in many utensils, which are treated with Teflon, or similar.

4.5 MEASUREMENTS OF CONTACT ANGLES AT LIQUID–SOLID INTERFACES

The magnitude of the contact angle, θ, between a liquid and a solid can be determined by various methods. The method to be used depends on the system and the accuracy required. There are two most common methods: one using a direct microscope and a goniometer, another using photography (and digital analyses). It should be mentioned that the liquid drop, which one generally uses in such measurements, is very small, between 10 and 100 μL. There are two different systems of interest: liquid–solid or liquid$_1$–solid–liquid$_2$. In the case of some industrial systems (such as oil recovery), one needs to determine θ at high pressures and high temperatures. In these systems, the value can be measured using photography. Recently, digital photography has also been used, since the data can be analyzed by computer programs.

It is useful to consider some general conclusions from these data (Table 4.2). One defines a solid surface as wetting if the magnitude of θ is less than 90°. However, a

TABLE 4.2
Contact Angles, θ, of Water on Different Solid Surfaces (25°C)

Solid	θ
Teflon (PTE)	108
Paraffin wax	110
Polyethylene	95
Graphite	86
AgI	70
Polystyrene	65
Glass	30
Mica	10

solid surface is designated as *nonwetting* if θ is greater than 90°. This is a practical and semiquantitative procedure. It is also seen that water, due to its hydrogen-bonding properties, exhibits a large θ on nonpolar surfaces (polyethylene terephthalate (PET)). On the other hand, one finds lower θ values on polar surfaces (glass, mica).

However, in some applications, one may change the surface properties by chemical modifications of the surface. For instance, PS has some weak polar groups at the surface. If one treats the surface with H_2SO_4, which forms sulfonic groups (Birdi, 1981), this leads to values of θ lower than 30° (depending on the time of contact between sulfuric acid and PS surface). PS petri dishes are used for growing bacteria. However, bacteria (which are generally negatively charged) do not attach to the PS (it being neutral in charge). After treatment with sulfuric acid (or similar), bacteria are found to attach to the dishes, and grow for investigation. This treatment (or similar) has been used in many other applications where the solid surface is modified to achieve a specific property. Since only the surface layer (a few molecules deep) is modified, the solid properties of the bulk do not change. This analysis shows the significant role of studying the contact angle of surfaces in relation to the application characteristics.

The magnitude of the contact angle of water (for example) is found to vary depending on the nature of the solid surface. The magnitude of θ is almost 100° on a waxed surface of a car paint. The industry strives to create such surfaces to give θ > 150°, the so-called super-hydrophobic surfaces. The large θ value means that water drops do not wet the car polish and are easily blown off by wind. The car polish is also designed to leave a highly *smooth* surface. In many industrial applications, one is concerned with both smooth and rough surfaces. The analyses of θ on *rough surfaces* will be somewhat more complicated than on smooth surfaces. The liquid drop on a rough surface (Figure 4.4) may show the *real* θ, real value (solid line), or some lower value (*apparent*) (dotted line), depending on the orientation of the drop.

However, no matter how rough the surface, the forces will be the same as those between a solid and a liquid. The contact angle is a thermodynamic quantity, and only related to the surface tension force equilibrium. The state or degree of roughness has no relation to this equilibrium state of surface forces (Birdi, 1993). The surface roughness may show contact angle *hysteresis* if one makes the drop move, but this will arise from other parameters (e.g., wetting and dewetting) (Birdi, 1993). Furthermore, in practice, the surface roughness is not easily defined. A *fractal* approach has been used to achieve a better understanding (Feder, 1988; Avnir, 1989; Coppens, 2001; Birdi, 1993; Yu and Li, 2001).

Even though Young's equation may sometimes not be easily applicable in some systems, many useful conclusions can be obtained if systematic data are available. In fact, currently, much key research in industry is based on Young's equation analyses. For example, in the data of Cos(θ), various liquids on Teflon gave an almost straight line plot, with the following relationship:

$$Cos(\theta) = k_1 - k_2 \gamma_L \tag{4.11}$$

This can also be rewritten as

$$Cos(\theta) = 1 - k_3 (\gamma_L - \gamma_{cr}) \tag{4.12}$$

TABLE 4.3

Some Typical γ_{cr} Values for Solid Surfaces

Surface Group	γ_{cr}
$-CF_2-$	18
$-CH_2-CH_3-$	22
Phenyl-	30
Alkyl chloride-	35
Alkyl hydroxyl	40

Source: Ross, 1963; Birdi, 2002.

where γ_{cr} is the critical value of γ_L at $Cos(\theta)$ equal to 0. The values of γ_{cr} have been reported for different solids using this procedure (Adamson and Gast, 1997). The magnitude of γ_{cr} for Teflon of 18 mN/m thus suggests that $-CF_2-$ groups exhibit this low surface tension. The value of γ_{cr} for $-CH_2-CH_3-$ alkyl chains gave a higher value of 22 mN/m than for Teflon. Indeed, from experience, one also finds that Teflon is a better water-repellent surface than any other material. The magnitudes of γ_{cr} for different surfaces are found to provide much useful information about the surface forces and the molecular structures (Table 4.3).

These data (Table 4.3) show the surface characteristics as related to γ_{cr}. In many cases, the surface of a solid may not behave as desired, and therefore the surface is treated accordingly, which results in a change of the contact angle of fluids.

4.6 THEORY OF ADHESIVES AND ADHESION

Adhesives are used in everyday applications. Adhesives may be in liquid form or thick pastes. The main mechanism is based on the polymerization or cross-linking of polymers, which gives rise to glue or other adhesive applications. Furthermore, if a liquid is removed from the surface of a solid, there will be work needed, W_{ad}, per square centimeter of solid exposed (Trevena, 1975; Adamson and Gast, 1997; Birdi, 2016). This process will require the destruction of 1 cm² of interface γ_{SL}, and the creation of 1 cm² of γ_S and γ_L. From this, one gets

$$W_{ad} = \gamma_S + \gamma_L - \gamma_{SL} \qquad (4.13)$$

which, on combining with Young's equation, gives

$$W_{ad} = \gamma_L \left(1 + Cos(\theta)\right) \qquad (4.14)$$

This shows that to remove a liquid in contact with a solid surface, the work needed is dependent on the surface tension and the contact angle, θ. If liquid wets the solid surfaces (e.g., water on glass $\theta < 90°$),

$$W_{ad} = 2\gamma_L \qquad (4.15)$$

W_{ad} is thus the work needed to create twice the surface tension (one on each side after separation or break up).

The wetting characteristics of *hydraulic fracking* fluid and shale will thus be of primary interest. This phenomenon is being investigated in relation to the degree of fluid (mostly water) recovery in the hydraulic fracking technology (Appendix II).

4.7 ADSORPTION/DESORPTION (OF GASES AND SOLUTES FROM SOLUTIONS) ON SOLID SURFACES (SHALE GAS RESERVOIRS)

All solid surfaces exhibit specific adsorption and desorption characteristics from their surroundings (gas, solutes from solutions).

For example, in the case of a shale reservoir, the adsorption and desorption of gas, such as methane (CH_4), will be of major interest as regards the fracking characteristics and gas recovery (desorption) (Figure 4.5). Shale gas is predominantly found in fine-grained, organic-rich (kerogen) rocks that are sometimes considered to be source reservoirs. It is found that these are rocks where gas is tightly bound and very little gas has been released. The shale is thus the source rock for natural gas. In the shale gas reservoir recovery process, the adsorption and desorption of gas (such as methane [CH_4]) is of major interest (Figure 4.5). The primary objective is to recover the adsorbed gas (mostly methane) (i.e., desorption of gas) from the shale matter (Chattoraj and Birdi, 1984; Ross and Bustin, 2007; Javadpour, 2009; Kale et al., 2010; Ambrose et al., 2011, 2012; Shabro et al., 2012; Birdi, 1997, 2016; Kumar, 2011; Fengpeng et al., 2014; Wu et al., 1993; Theodori et al., 2014; Vafai, 2015; Zelenev, 2011; Yoon et al., 1990). The gas molecules are most likely present where the plants (organic material; kerogen) had transformed into CH_4, and so on. The methane molecules are therefore expected to be finely dispersed (at high pressure and high temperature). It is therefore useful to understand the surface adsorption–desorption forces under such conditions (gas–solid interface) (Kumar, 2011). Also, the solid (i.e., shale) is a complex rock, not a well-defined or uniform material (Calvin, 1969; Donaldson, 2013).

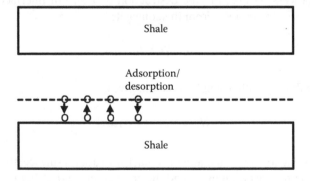

FIGURE 4.5   Shale gas reservoir (adsorption–desorption of gas).

Hence, the gas-recovery rate will be found to vary from shale source to source (since the composition of shale is complex and variable). Microporous rocks in particular, such as those found in shale reservoirs, are now considered as important energy sources. This is indeed also observed in the reservoir recovery processes. Shale rock is known to be of very low porosity (Appendix I). The gas molecules after desorption have to move through very narrow pores (of molecular size). The diffusion mechanism has been analyzed based on the size of the pores. The process of gas diffusion through narrow (of molecular size) pores has been investigated for many decades. This process is explained in recent studies to be the same as the Knudsen regime (Knudsen, 1952; Ruthven, 1984; Feres and Yablonsky, 2004; Bird, 2007; Freman et al., 2009; Allan and Mavko, 2013; Engelder et al., 2014). Model studies from microporous rocks have been based on the following parameters:

- Adsorbed monolayer
- Knudsen diffusion
- As a function of pressure

The pore size (and shape) distribution will also be expected to be different for different rocks. It is reported that permeability is about 10^{-18} to 10^{-21} m^2, and that the pore size ranges from a few nanometers to several microns. The latter value is slightly larger than the methane molecular mean free path. Recent studies have been reported where the shale gas recovery is described on the basis of models related to (Figure 4.6a and b):

$$\text{Diffusion—}\textit{adsorption/desorption—}\text{Darcy flow}$$

The diffusion process of a gas is related to the pore diameter (Figure 4.6b). The different gas diffusion mechanisms are as follows:

- Pore diameter/type of process
- 10^{-4} to 10^{-3} μm: molecular diffusion
- 10^{-3} to 10^{-2} μm: surface diffusion (lateral diffusion)
- 10^{-2} to 10^{-1} μm: Knudsen diffusion (mean-free path length)
- 10^{-1} to 10 μm: free diffusion

Since the pore size in shale is of wide range, it is thus composed of a combination of different diffusion processes. Various model flow analyses have been reported. The impact of molecular diffusion is considered to be of importance. The composition of gas changes during production over time. The Knudsen regime is considered to become important in some shale gas matrices. In classical theory on gas kinetics, the mean free path (L) is defined as the mean distance a molecule can travel before colliding with another gas molecule. The Knudsen number (K_N) is defined as the ratio of L and the diameter of the pore throat (d_P):

$$K_N = L / d_P$$

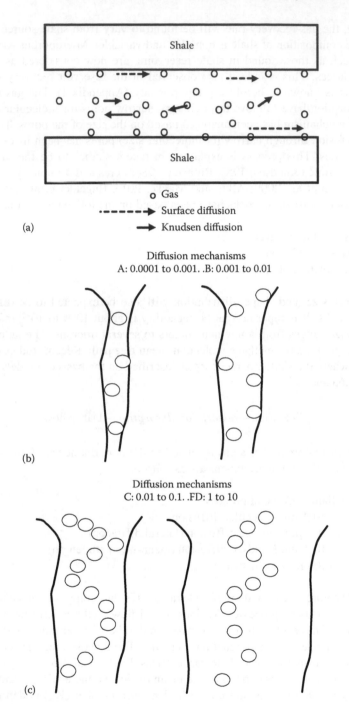

FIGURE 4.6   (a) A diffusion model of gas recovery in a shale reservoir. (b) Different transport mechanisms in porous solids (A) molecular diffusion, (B) surface diffusion, (C) Knudsen diffusion, and (FD) free diffusion (micrometer scale).

It is also found that Darcy's law is valid for systems with a K_N less than 0.001. The gas recovery is thus a series of specific surface-related steps. In a shale gas reservoir, therefore, one will have a system:

Solid phase (shale)—*Gas phase* (mostly methane)—Water phase

Further, the gas in a shale reservoir is found to be at a higher energy state than at the well hole. This is in accord with the fact that the shale gas has diffused to the reservoirs where conventional gas reservoirs are found. However, the gas shale reservoir is somewhat more complex. There are other parameters besides the desorption phenomenon, such as diffusion inside the pores (so-called Knudsen diffusion) (Thomas and Clouse, 1990; Freeman et al., 2009; Shabro et al., 2012; Fengpeng et al., 2014) (Figure 4.7). The latter is observed where gas diffusion takes place in pores of molecular dimensions. The solid surface interacts with gases or liquids as determined by various kinds of surface forces. The adsorption process on a solid surface is described in the following ways:

- Adsorption: Gas or vapor (adsorbate) adsorption on a solid surface (adsorbent)
- Adsorbent: Solid surface with a large area/weight (m^2/gram)
- Adsorbate: A substance that adsorbs on a solid surface
- Absorption: A process where a gas/vapor may diffuse into the porous solid
- Sorption: Sometimes this term is used for absorption

The adsorption energy of a nonpolar molecule, such as methane, will be mainly Van der Waal's type (which is known to be related to the critical temperature, T_c, of the gas) (Adamson and Gast, 1997; Chattoraj and Birdi, 1984; Birdi, 2016; Somasundaran, 2015). The adsorption of gas on a solid surface is very important in various systems, especially in industries involved with catalysis. In the case of shale gas reservoirs, it is found that the quantity of the TOC mainly determines the degree of adsorption (Ross and Bustin, 2007; Chalmers and Busstin, 2007). The molecules in gas are moving very fast (kinetic energy), but on adsorption, there will thus be a large decrease in kinetic energy (and thus a decrease in entropy, ΔS_{ad}) (Chattoraj and Birdi, 1984; Birdi, 2016; Somasundaran, 2015). Adsorption takes place spontaneously, which means

$$\Delta G_{ad} = \Delta H_{ad} - T\Delta S_{ad} \qquad (4.16)$$

that ΔG_{ad} is negative, which suggests that ΔH_{ad} is also negative (exothermic). The gas adsorption on a solid is accompanied by a decrease in entropy (i.e., $\Delta S_{ad} < 0$) (Chattoraj and Birdi, 1984; Auroux, 2013; Somorjai, 2000; Somasundaran, 2016). This is related to the fact that the degrees of freedom of the adsorbed molecules are lower than in the gaseous state. The adsorption of gas can be of different

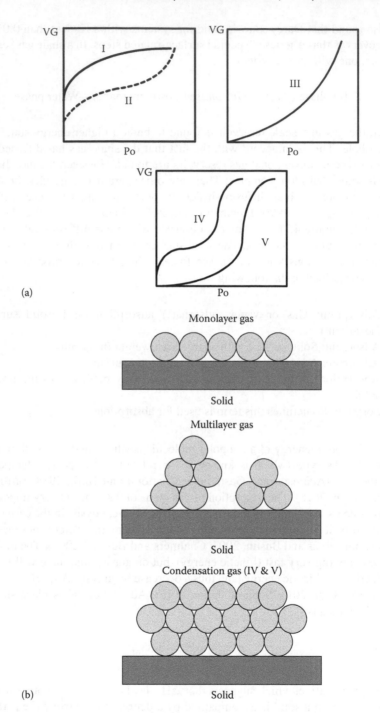

FIGURE 4.7 (a) Plot of v_{gas} vs. P_{gas} (Types I, II, III, IV, and V) (see text for details). (b) Molecular description of various gas-adsorption phenomena (Types I, II, III, IV, and V).

types. The gas molecule may adsorb as a kind of condensation process, or it may, under other circumstances, react with the solid surface (chemical adsorption or chemisorption). In the case of chemo-adsorption, one almost expects a chemical bond formation. On carbon, while oxygen adsorbs (or chemisorb), one can desorb CO or CO_2. The experimental data can provide information on the type of adsorption. This type of adsorption phenomenon will also be expected in shale (Figure 4.7). On porous solid surfaces, the adsorption may give rise to *capillary condensation*. This indicates that porous solid surfaces will exhibit some specific properties.

The adsorption processes one finds most in industry are catalytic reactions (e.g., the formation of NH_3 from N_2 and H_2).

The surface of a solid may differ in many ways from its bulk composition. In particular, such solids as commercial carbon black may contain minor amounts of impurities (such as aromatics, phenol, carboxylic acid). This would render surface adsorption characteristics different from pure carbon.

The composition of a silica surface has been considered to be O–Si–O, as well as hydroxyl groups formed after interaction (hydrolysis) with water molecules. The orientation of the different groups may also be different at the surface. As another example, carbon black has been reported to possess different kinds of chemical groups on the surface (Birdi, 2009). These different groups are aromatics, phenol, carboxylic, and so on. These different sites on carbon can be estimated by comparing the adsorption characteristics of different adsorbents (such as hexane and toluene). This arises from the fact that on all solid surfaces the adsorption is selective.

When any clean solid surface is exposed to a gas, the latter may adsorb on the solid surface to varying degrees. Gas adsorption on solid surfaces may not stop at a monolayer state. More than one layer (*multilayer*) adsorption will take place only if the pressure is reasonably high. Experimental data show this when the volume of gas adsorbed, v_{gas}, is plotted against P_{gas} (Figure 4.7a).

These analyses show that there are five different kinds of adsorption states (Figure 4.7a and b). The data of adsorption isotherms were classified based on v_{gas} versus P_{gas} (Schedulko, 1966; Adamson and Gast, 1997; Chattoraj and Birdi, 1984; Kumar, 2012). The various adsorption states were delineated as

- Type I: These are obtained for Langmuir adsorption
- Type II: This is the most common type where multilayer surface adsorption is observed
- Type III: This is somewhat special type with almost only multilayer formation, such as nitrogen adsorption on ice
- Type IV: If the solid surface is porous then this is found similar to Type II
- Type V: On porous solid surfaces Type III

These adsorption isotherms were found to be related to the molecular packing on the solid surface as shown in Figure 4.7b. Type I is a monolayer structure. Type II is the multilayer packing, which is found to be the most common phenomenon. Type

III is not very common, since it is a process related to the enthalpy of adsorption and condensation as follows:

$$\text{Heat of adsorption} \leq \text{heat of condensation}$$

Types IV and V of isotherms are generally found on porous solids. This suggests capillary condensation phenomena at work (Appendix III).

As regards the pores in a porous solid surface, it is found that these vary from 2 to 50 nm. This aspect is classified as follows.

- *Micropores* are in the range of 2 nm.
- *Macropores* are designated for larger than 50 nm.
- *Mesopores* are used for 2–50 nm range.

The Brunauer–Emmett–Teller (BET) model suggests that the molecular arrangement on the solid surface can be estimated from the molecular size and geometrical arrangement of the adsorbate. The spacing of the adsorbed gas molecules will be related to the adsorption potential. This will vary on different solids. The BET monolayer packing may approximate to either the bulk liquid or the solid state. The molecular dimension (A_m) can be estimated as (Emmett and Brunauer, 1937; Corrin, 1951; Adamson and Gast, 1997):

$$A_m = f_g \left(M/d_d N_A \right)^{2/3} \times 10^{14} \, nm^2 / \text{molecule} \qquad (4.17)$$

where:
M is the molecular weight of the gas
d_d is the density of the bulk phase
N_A is the Avogadro number
f_g is the packing factor (which for hexagonal packing is equal to 1.091)

It was suggested that, in the adsorbed monolayer, atoms can be closer packed than in the bulk phase (liquid or solid). The experimental and calculated values of A_m are given in Table 4.4.

TABLE 4.4
Experimental and Calculated (Liquid Density) Values of A_m for Various Gases

Gas	Temperature/K	A_m (nm²/ Mmolecule) Equation 4.17	Experimental
N_2	77.5	0.162	0.162
CO	77.5	0.160	0.147
C_5H_{12}	293	0.362	0.523
C_6H_{14}	273	0.390	0.589

These data indicate that some molecular geometrical packing arrangement exists, as assumed in Equation 4.17. However, in the case of nonspherical molecules (C_5H_{12} and C_6H_{14}), the experimental data indicate larger spacing than in the liquid phase.

4.7.1 GAS ADSORPTION ON SOLID MEASUREMENT METHODS

The mechanism of gas adsorption on a solid has been measured by using different experimental methods (Adamson and Gast, 1997; Kumar, 2012; Somasundaran, 2014; Birdi, 2014). Since the gas–solid system is most important in the catalysis industry, there have been great advances in our knowledge on the subject. In the following, some of these different methods are delineated.

4.7.1.1 Gas Volumetric Change Methods of Adsorption on Solids

The change in the volume of gas during adsorption is measured directly in principle, and the apparatus is comparatively simple (Figure 4.8).

One can use a mercury (or similar kind of liquid) reservoir beneath the manometer, and the burette to control the levels of liquid in the apparatus above. Calibration involves measuring the volumes of the gas (v_g) lines and of the void space (Figure 4.8). All pressure measurements are made with the right arm of the manometer set at a fixed zero point so that the volume of the gas lines does not change when the pressure changes. The apparatus, including the sample, is evacuated and the sample heated to remove any previously adsorbed gas. A gas such as helium is usually used for the calibration, since it exhibits very low adsorption on the solid surface. After helium is pushed into the apparatus, a change in volume is used to calibrate the apparatus and the corresponding change in pressure is measured. A different gas (such as nitrogen) is normally used as the adsorbate if one needs to estimate the surface area of a solid. The gas is cooled by liquid nitrogen. The tap to the sample bulb is opened, and the drop in pressure is determined. In the surface area calculations, one uses a value of 0.162 nm² for the area of an adsorbed nitrogen molecule. Due to the toxic properties of Hg, modern apparatuses

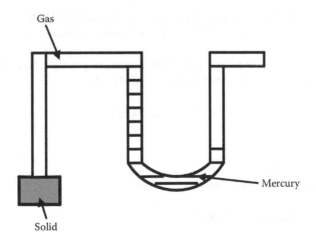

Gas

Mercury

Solid

FIGURE 4.8 Gas adsorption on solid apparatus.

use a combination of valves to measure the change in the volume of gas adsorbed. Commercially available instruments are designed with such modern detectors.

4.7.1.2 Gravimetric Gas Adsorption Methods

The amount of gas adsorbed on the solid surface is generally very small. A modern sensitive microbalance is used to measure the adsorption isotherm. The sensitivity is very high since only the change in weight is measured. These microbalances can measure weight differences in nanograms, micrograms, and smaller. With such extreme sensitivity, it is possible to measure the weight change caused by the adsorption of a *single monolayer* on a solid if the surface area is large. The normal procedure is to expose the sample to the adsorbate gas at a certain pressure, allowing sufficient time for equilibrium to be reached, and then determining the mass change. This is repeated for a number of different pressures, and the number of moles adsorbed as a function of pressure is plotted to give an adsorption isotherm. Microbalances (stainless steel) can be made to handle pressures as high as 120 Mpa (120 atm), since gases that adsorb weakly or boil at very low pressures can still be used.

4.7.1.3 Langmuir Gas Adsorption

Shale gas recovery has been analyzed by applying the Langmuir adsorption theory (Chattoraj and Birdi, 1984; Engelder, 2014; Shabro et al., 2014; Fengpeng et al., 2014). The gas recovery (mainly a desorption and diffusion process) is known to be a complicated system. Currently, no standard or viable model exists to predict gas production in the shale formations (Passey et al., 2010; Lu et al., 1995; Gault and Stotts, 2007; Javadpour, 2009; Sondergeld et al., 2010; Ambrose et al., 2010; Kale et al., 2010; Freeman et al., 2010; Swami and Settari, 2012; Zhai et al., 2014). According to the Langmuir model (Adamson and Gast, 1997; Birdi, 2016; Shabro et al., 2012), an assumption is made that only one layer of gas molecule adsorbs. A monolayer of gas adsorbs in the case where there are only a given number of adsorption sites. This can be expressed as follows:

Gas molecule in the bulk phase + Active site on the adsorbent surface
= Localized adsorption of the gas–solid

This is the simplest adsorption model. The amount of gas adsorbed, N_s is related to the monolayer coverage, N_{sm}, as follows (Appendix II):

$$N_s / N_{sm} = \mathbf{a}p / (1 + \mathbf{a}p) \tag{4.18}$$

where:
- p is the pressure
- $\mathbf{a}$ is dependent on the energy of adsorption

This equation can be rearranged as

$$p / N_s = \left(1 / \left(\mathbf{a}N_{sm}\right) + p / N_{sm}\right) \tag{4.19}$$

A plot of p/N_s versus p, of the data are used to analyze the system. The plot will be linear (in general) and the slope is equal to $1/N_{sm}$. The intersection gives the magnitude of **a**. The equilibrium state is controlled by the pressure and the chemical potential energy of the surface and the gas.

It is useful to consider a few examples of gas adsorption data. Charcoal is found to adsorb 15 mg of N_2 as a monolayer. Another example is that of the adsorption of N_2 on a mica surface (at 90°K). The following data were found:

Pressure/Pa	Volume of Gas Adsorbed (at STP)
0.3	12
0.5	17
1.0	24

In Equation 4.15, one assumes

- That the molecules adsorb on definite sites
- That the adsorbed molecules are stable after adsorption

It is clear that the amount of gas adsorbed increases with pressure. In this system, one finds that the relation between pressure and amount adsorbed is

Pressure increase $= 1.0/0.3 = 0.33$
Volume of gas adsorbed increase $= 24/12 = 2$

This means that in gas reservoirs, when the pressure drops (after the fracturing step), then gas will be desorbed, as has also been found in production wells. Furthermore, the magnitude of the surface area of the solid can be estimated from the plot of p/N_s versus p. Most data fit this equation under normal conditions, and it is therefore widely applied to analyze the adsorption process. Langmuir adsorption data for nitrogen on mica (at 90°K) were found to be

$p = 1$/Pa	2/Pa
$Vs = 24$ mm^3	28 mm^3

This shows that the amount of gas adsorbed increases by a factor $28/24 = 1.2$ when the gas pressure increases twofold.

4.7.2 Various Gas Adsorption Analyses

In combination with the Langmuir equation, one can derive the following relation between N_s and p:

$$N_s = Kp \qquad (4.20)$$

where **K** is a constant. This is the well-known Henry's law relation, and it is found to be valid for most isotherms at low relative pressures. In those situations where the ideal (Equation 4.16) does not fit the data, the Van der Waals equation type of corrections have been suggested.

The *adsorption–desorption* process is of interest in many systems (such as cement, oil/gas reservoirs, underground water flow [pollution], etc.). The water vapor may condense in the pores after adsorption under certain conditions. This may be studied by analyzing the adsorption–desorption data (Figure 4.9).

Multilayer gas adsorption: In some systems, the adsorption of gas molecules proceeds to higher levels where *multilayers* are observed. The adsorption isotherm of Type I corresponds to the Langmuir equation. These are found from data analyses where one finds that multilayer adsorption takes place (Figure 4.10).

Type II isotherm indicates multilayer adsorption with high adsorption potential. This type is observed on a surface of high adsorption potential. Type III is indicative of low adsorption potential. This type is comparatively uncommonly observed on porous solid surfaces. It has been observed on the following systems:

- Bromine adsorption on silica gel
- Adsorption of nitrogen on ice

Types IV and V isotherms are observed on porous solids. The modified BET equation, which has been derived for multilayer adsorption data, can provide useful

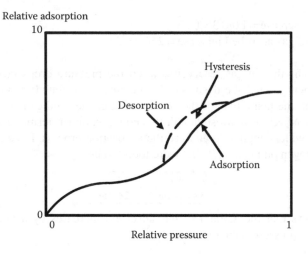

FIGURE 4.9   Adsorption (N_s/N_{sm} = relative adsorption) vs. pressure ($p/p°$ = relative pressure) of a gas on solid.

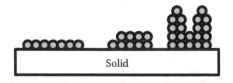

FIGURE 4.10   BET model for multilayer adsorption on solids.

information. The enthalpy involved in multilayers is related to the differences in adsorption process, and was defined by BET theory as

$$E_{BET} = \exp\left[\left(E_1 - E_v\right)/RT\right] \qquad (4.21)$$

where E_1 and E_v are enthalpies of desorption. The BET equation thus, after modification of the Langmuir equation, becomes

$$p/\left(N_s\left(p^\circ - p\right)\right) = 1/E_{BET}N_{sm} + \left[\left(E_{BET} - 1\right)/\left(E_{BET}N_{sm}\right)\left(p/p^\circ\right)\right] \qquad (4.22)$$

A plot of the adsorption data of the left-hand side of this equation versus relative pressure (p/p°) allows one to estimate N_{sm} and E_{BET}. The magnitude of E_{BET} is found to give data plots that are either Type III or Type II. Experiments show that when the value of E_{BET} is low, which means that the interaction between the adsorbate and the solid is weak, then Type III plots are observed.

4.7.3 ADSORPTION OF SOLUTES FROM SOLUTION ON SOLID SURFACES

Another system, that is, the adsorption of a solute from a solution on a solid surface, is of much importance in everyday life, and relates to water purification, oil spills in oceans, industry wastewater, hydraulic fracking technology, and so on.

A clean solid surface is actually an active center for adsorption from the surroundings, for example, air or liquid. Experiments show that a perfectly cleaned metal surface, when exposed to air, will adsorb a single layer of oxygen or nitrogen (or water) from its surroundings (Freundlich, 1926; Birdi, 2016). Or, when a completely dry glass surface is exposed to air (with some moisture), the surface will adsorb a monolayer of water (moisture). In other words, the solid surface is not as inert as it may seem to the naked eye. This has many consequences in industry, such as corrosion control, and oil and gas reservoirs. Additionally, in hydro-fracking, the adsorption of additives on solid (shale, etc.) surfaces is important to the process. It is reported that some molecules (such as alcohols and surface-active substances) are related to fracture formation in rocks (Rehbinder and Schukin, 1972). As mentioned elsewhere (Appendix II), one finds that in fracking fluids, alcohols (or other similar surface active fracturing substances (SFAFS)) are being used in shale reservoir technology.

Adsorption at the liquid surfaces of solutes can be analyzed using the Gibbs adsorption isotherm since the surface energy γ of a solution can be readily measured (Chattoraj and Birdi, 1984; Adamson and Gast, 1997; Birdi, 2003, 2016; Freundlich, 1926). However, for solid substrates, this is not the case, and the adsorption density has to be measured in some other manner. In the present case, the concentration of an adsorbate in solution will be monitored. In place of the Gibbs equation, one can use a simple adsorption model based on the mass action equilibrium approach. On any solid surface, a certain number of possible adsorption sites per gram (N_m) will be expected. This is the number of sites where any adsorbate can freely adsorb. There, a fraction, θ, will be filled by one adsorbing solute. It will also be expected that an *adsorption–desorption* process will exist at the surface. The rate of *adsorption* will be given as

$$(\text{concentration of solute})(1-\theta)N_m \tag{4.23}$$

and the rate of *desorption* will be given as

$$(\text{concentration of solute})(\theta)N_m \tag{4.24}$$

It is known that, at equilibrium, these rates must be equal:

$$k_{ads}C_{bulk}(1-\theta)\ N_m = k_{des}\theta N_m \tag{4.25}$$

where:

k_{ads}, k_{des} are the respective proportionality constants
C_{bulk} is the bulk solution concentration of solute

The equilibrium constant, $K_{eq}=k_{ads}/k_{des}$, which gives

$$C_{bulk}/\theta = C_{bulk}+1/K_{eq} \tag{4.26}$$

and since $\theta=N/N_m$, where N is the number of solute molecules adsorbed per gram of solid, one can write

$$C_{bulk}/N = C_{bulk}/N_m+1/(K_{eq}N_m) \tag{4.27}$$

Thus, measurement of N for a range of concentrations (C) should give a linear plot of C_{bulk}/N against C_{bulk}, where the slope gives the value of N_m and the intercept the value of the equilibrium constant K_{eq}. This model of adsorption is referred to as the *Langmuir adsorption isotherm*. The aim of this experiment is to test the validity of this isotherm equation and to measure the surface area per gram of charcoal, which can easily be obtained from the measured N_m value, if the area per solute molecule is known. In a typical experiment, adsorption experiments are carried out as follows. The solid sample (e.g., activated charcoal) is shaken in contact with a solution with a known concentration of solute. After equilibrium is reached (typically after 24 h), the amount of solute adsorbed is determined by a suitable analytical method. One may also use solutions of dyes (such as methylene blue), and, after adsorption, the amount of dye in the solution is measured by any convenient spectroscopic method (Raman, VIS, UV, or fluorescence).

4.7.4 Solid Surface Area (Area/Gram) Determination

In all applications where finely divided powders are used (such as talcum powder or charcoal powder), the property of these will depend mainly on the *surface area per gram* (varying from a few square meters (talcum powder) to over 1000 m²/g (charcoal)). For example, if one needs to use charcoal to remove some chemical (such as coloring substances) from wastewater, then it is necessary to know the amount of absorbent needed to complete the process. In other words, if one needs 1000 m² for the adsorption when using charcoal, then 1 g of solid will be required. In fact, under normal conditions, if one swallows charcoal it is considered dangerous, because it leads to the removal of essential substances from the stomach lining (such as lipids and proteins).

The estimation of the surface area of finely divided solid particles from solution adsorption studies is subject to many of the same considerations as gas adsorption, but with the added complication that larger molecules are involved, whose surface orientation and pore penetrability may be uncertain. A first condition is that a definite adsorption model is obeyed, which, in practice, means that area determinations are limited to cases in which the simple Langmuir equation (Equation 4.26) is valid. The constant rate is found, for example, from a plot of the data according to Equation 4.26, and the specific surface area then follows from Equation 4.27. The surface area of the adsorbent is generally found easily from the literature (such as molecular models).

In the case of gas adsorption where the BET method is used, it is reasonable to use the Van der Waals area of the adsorbate molecule. In the case of being small, or even monoatomic, surface orientation is not a major problem. In the case of adsorption from solution, however, the adsorption may be chemisorption.

In the literature, fatty acid adsorption for surface area estimation has also been used. This is useful since fatty acids are known to pack perpendicular to the surface (self-assembly, monolayer formation), and with the close-packed area per molecule of about 20.5 Å². In all of these cases, the adsorption is probably chemisorption, involving hydrogen bonding or actual salt formation with surface oxygen. If polar solvents are used to avoid multilayer formation on top of the first layer, the apparent area obtained may vary with the solvent used. In the case of stearic acid on a graphitized carbon surface, graphon, the adsorption, while obeying the Langmuir equation, appears to be physical, with the molecules oriented flat on the surface.

In fracking fluids, one uses a variety of additives (Appendix II). Knowledge of the adsorption characteristics of these substances in the shale reservoirs is important. As another example, the adsorption of surfactants on polycarbonate indicated that, depending on the surfactant and the concentration, the adsorbed molecules might be lying flat on the surface perpendicular to it, or they might form a bilayer. The mechanism of adsorption is easily determined from the experimental data. In the case of bilayer adsorption, one finds charge-reversal.

A second class of adsorbates, of which much use has been made, is that which is based on the use of dyes. This method has the advantage that it is easy to measure the concentration of dye (less than ppm). The adsorption equilibrium generally follows the Langmuir equation, to be multilayer. For example, the apparent molecular area of 19.7 Å² for methane blue on graphon or larger than the actual molecular area of 17.5 Å², but apparent value for the more oxidized surface of spheron was about 10.5 Å² molecule. However, this procedure assumes that the dye's molecules are present as monomers in the solution. The fatty acid adsorption method has been used by many investigators. In some instances one has also used pyridine (as an adsorbate) for solid oxide surface studies. The adsorption data followed the Langmuir equation. The effective molecular area of pyridine is about 24 Å^2 per molecule.

In the literature, many different approaches have been proposed to estimate the surface area of a solid. The magnitudes of surface areas may be estimated from the exclusion of like-charged ions from a charged interface. This method is intriguing in that no estimation of either site or molecular area is needed. In general, however, surface area determination by means of solution adsorption studies, while convenient experimentally, may not provide the most correct information. Nonetheless, if a solution adsorption procedure has been standardized for a given system, by means of independent checks, it can be very useful in determining the relative areas of a series of similar materials. In all cases, it is also more real as it is what happens in real life.

Experimental Method for Adsorption Analysis: A typical procedure was to use 1.0 g of alumina powder and add 10 mL of detergent solution with varying concentrations. The mixture was shaken and the concentration of the detergent was estimated by some suitable method. It was found that equilibrium was obtained after 2–4 h. The detergent, such as dodecyl-ammonium chloride (DAC), was found to adsorb 0.433 mM/g of alumina with a surface area of 55 m²/g. The surface area of alumina as determined from stearic acid adsorption (and using the area/molecule of 21 Å^2 from monolayer), gave a value of 55 m²/g. These data can be analyzed in more detail.

$$\text{Surface area} = 55\,\text{m}^2/\text{g}$$

$$\text{Amount adsorbed} = 0.433\,\text{mM/g}$$

$$= 0.433 \times 10^{-3}\,6 \times 10^{23}\,\text{molecules}$$

$$= 0.433 \times 10^{20}\,\text{molecules}$$

$$\text{Area/molecule of DAC} = 55 \times 10^4 \times 10^{16}\,\text{Å}^2 / 0.433 \times 10^{20}$$

$$= 55 / 0.433\,\text{Å}^2$$

$$= 127\,\text{Å}^2$$

This value is in keeping with data from other experiments. The adsorption isotherms obtained for various detergents showed a characteristic feature that an equilibrium value was obtained when the concentration of detergent was over the critical micelle concentration (CMC). The adsorption was higher at 40°C than at 20°C. However, the shapes of the adsorption curves were the same (Birdi, 2003). One can also calculate the amount of a small molecule, such as pyridine (mol. wt. 100), adsorbed as a monolayer on charcoal with 1000 m²/g. In the following, these data are delineated:

Area per pyridine molecule $= 24 \text{ Å}^2 = 24 \times 10^{-16} \text{ cm}^2$
Surface area of 1 g charcoal $= 1000 \text{ m}^2 = 1000 \times 10^4 \text{ cm}^2$
Molecules of pyridine adsorbed $= 1000 \times 10^4 \text{ cm}^2 / 24 \times 10^{-16} \text{ cm}^2 / \text{molecule}$
$= 40 \times 10^{20} \text{ molecules}$
Amount of pyridine/g charcoal adsorbed $= (40 \times 10^{20} \text{ molecules}/6 \times 10^{23})100$
$= 0.7 \text{ g}$

This is a useful example to illustrate the application of charcoal (or similar substances with large surface area per gram) in the removal of contaminants by adsorption.

4.8   SURFACE PHENOMENA IN SOLID-ADSORPTION AND FRACTURE PROCESS (BASICS OF FRACTURE FORMATION)

Any solid matter will crack when sufficient force is applied. The mechanism of the breakdown of a solid structure has been investigated for many decades. A simple description of this relatively complicated phenomenon is delineated here. A special example is considered: a solid immersed in a liquid phase (solid immersed in a water solution). A solid immersed in a water solution will adsorb solute, as delineated above. In the literature, studies have been reported regarding the effect of liquids with surface-active properties on the fracture process when suitable pressure is applied (El-Shall and Somasundaran, 1984). The minimum stress, S_{solid}, required for the fracture of a solid is described as

$$S_{solid} = \left(4 \, E_Y \gamma / L_{fracture} \right) 1 / 2 \tag{4.28}$$

where:

E_Y	is Young's modulus
γ	is the free energy of the created surface of the fracture
$L_{fracture}$	is the crack (fracture) length

Various studies showed that surface-active fracture substances (SAFS) had an effect on the stress energy of fracture formation (by decreasing the magnitude of γ). The effect of SAFS (ethylene glycol) on cement was reported (Gupta and Hildek, 2009; Ma and Holditch, 2016). In hydraulic fracking, liquids with different SAFS have been used, which are related in molecular structure to the latter.

4.9   HEATS OF ADSORPTION (DIFFERENT SUBSTANCES) ON SOLID SURFACES

In order to understand the mechanism of any process, one needs to know the heat of the reaction. A solid surface interacts with its surrounding molecules (in gas or liquid phase) with varying degrees. For example, if a solid is immersed in a liquid, the interaction between the two bodies will be of interest (Figure 4.11). The interaction of a substance with a solid surface can be studied by measuring the heat of

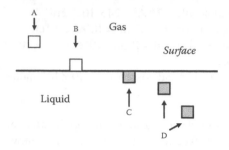

FIGURE 4.11   Solid immersion process in a liquid.

adsorption (besides other methods). The information one needs is whether the process is exothermic (heat is produced) or endothermic (heat is absorbed). This leads to an understanding of the mechanism of adsorption and helps in the application and design of a system. Calorimetric measurements have provided much useful information. When a solid is immersed in a liquid, one finds that, in most cases, there is a liberation of heat (Figure 4.11):

$$q_{imm} = E_S - E_{SL} \qquad (4.29)$$

where:
    E_S is the surface energy of the solid surface
    E_{SL} is the surface energy of the solid surface in liquid

The quantity, q_{imm}, is measured from calorimetry where temperature change is measured after a solid (in a finely divided sate) is immersed in a given liquid. Since these measurements can be carried out with micro-calorimetric sensitivity, much systematic data are reported in the literature.

One will expect that when a polar solid surface is immersed in a polar liquid, there will be larger q_{imm} than if the liquid was an alkane (nonpolar). Values of some typical systems are depicted in Table 4.4.

These studies (Table 4.5) show that adsorption mechanisms can be studied from heat of immersion data. Modern calorimeters with very high sensitivity have provided much useful information on various solid–liquid systems.

TABLE 4.5

Heat of Immersion (q_{imm}) (erg/cm² at 25°C)

Solid	Liquid	Polar (H_2O–C_2H_5OH)	Nonpolar (C_6H_{14})
Polar	(TiO_2; Al_2O_3; glass)	400–600	100
Nonpolar	(graphon; Teflon)	6–30	50–100

4.10   SOLID SURFACE ROUGHNESS (DEGREE OF SURFACE ROUGHNESS)

The nature of any solid surface (surface area, surface roughness) plays an important role in many applications (Adamson and Gast, 1997; Chattoraj and Birdi, 1984). In fact, in some applications it is the main criterion. For instance, friction decreases appreciably as the surface of a solid becomes smooth. The shine of a solid surface is enhanced as it is finely polished. This arises from the fact that the number of surface molecules that are able to come in contact with another solid or liquid phase is reduced. Thus, in some cases, one prefers roughness (high friction) (roads, shoe soles) while in other systems (glass, office table) one requires smooth, solid surface characteristics. The solid surfaces that one finds are manufactured by different methods: sawn, cut, turned, polished, or chemically treated. All of these procedures leave the solid surface rough, to varying degrees. In industry, one finds various methods, which can characterize the roughness. Polishing is also an important application in the surface chemistry of solids. The surface layer produced after polishing may or may not remain stable after exposure to its surroundings (air, other gases, oxidation). The polishing industry is much dependent on the surface's molecular behavior. Catalysis technology is another important area where the solid surface is of primary importance (Chattoraj and Birdi, 1984; Adamson and Gast, 1997).

4.11   FRICTION (BETWEEN $SOLID_1$–$SOLID_2$)

Friction is defined as the resistance to sliding between two bodies (Adamson and Gast, 1997). Friction is lowered by using a suitable lubricant (specific for each system). In the case where the solids are very close, then the surface roughness becomes the determining factor. That means the resistance is higher between two rough surfaces. The degree of plasticity or deformation of the solids will also affect the friction; a lubricant will also reduce resistance if its viscosity is high. Thus, when one solid is sliding or rubbing against another, there are a variety of parameters. These parameters are also termed *tribology* (related to rubbing) (Adamson and Gast, 1997). The *coefficient of friction* can be appreciably reduced if boundary lubrication decreases the force field. This may be achieved by adsorbed films (Adamson and Gast, 1997).

4.12   PHENOMENA OF FLOTATION (OF SOLID PARTICLES TO LIQUID SURFACE) (WASTEWATER—HYDRAULIC FRACKING)

Solid–liquid interfacial forces have also been used in the selective separation of solids from water suspensions (such as wastewater systems, or processes for separating different minerals). Only in rare cases does one find minerals or metals in pure form (such as gold!). The earth's surface consists of a variety of minerals (major components are iron, silica, oxides, calcium, magnesium, aluminum, chromium, cobalt, titanium). Minerals found in nature are always mixed (e.g., zinc sulfide and feldspar). In order to separate zinc sulfide, one suspends the mixture in water, and air bubbles are made to achieve separation. This process is called

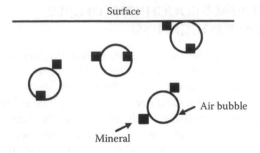

FIGURE 4.12 Flotation of mineral (or other material) particles as aided by air bubbles.

flotation (ore, which is heavier than water, is floated by bubbles). Flotation is a technical process in which suspended particles are clarified by allowing them to float to the surface of the liquid medium (Klimpel, 1995; Fuerstenau et al., 1985; Glembotskii et al., 1972; Kawartra, 1995; Klassen and Mokrousov, 1963; Rubinstein, 1995; Yoon and Luttrell, 1986; Adamson and Gast, 1997; Fuerstenau et al., 2007; Leja, 2012).

The material can thus be removed by skimming the surface. Economically, this is much cheaper than any other process. If the suspended particles (such as minerals) are heavier than the liquid, then one uses gas (air or CO_2, or other suitable gas) bubbles to enhance the flotation.

The flotation efficiency is also found to be dependent on the bubble size and size distribution. This observation thus relates the efficiency to the interfacial tension (IFT). In some cases, the desired mineral is rendered hydrophobic by the addition of a surfactant or collector chemical; the particular chemical depends on the mineral that is being refined. As an example, pine oil is used to extract copper. This slurry (more properly called the *pulp*) of hydrophobic, mineral-bearing ore and hydrophilic gangue is then introduced to a water bath, which is aerated, creating bubbles. The hydrophobic grains of mineral-bearing ore escape the water by attaching to the air bubbles, which rise to the surface, forming a foam (more properly called a *froth*). The froth is removed, and the concentrated mineral is further refined. The flotation industry is a very important area in metallurgy and other related processes. The flotation method is based on treating a suspension of minerals (ranging in size from 10 to 50 μm) in a water phase to air (or some other gas) bubbles (Figure 4.12).

Flotation leads to the separation of ores from their mixtures. It has been suggested that among other surface forces, the contact angle plays an important role. The gas (air, or other gas) bubble as attached to the solid particle should have a large contact angle for separation. Furthermore, it should be stable at the surface. Bubbles, as needed for flotation, are created by various methods, such as

- Air injection
- Electrolytic methods
- Vacuum activation

For example:

In a simple laboratory experiment (Adamson and Gast, 1997), one may use the following recipe: To a 1% sodium bicarbonate solution, one can add a few grams of sand. Then, if one adds some acetic acid (vinegar), the bubbles of CO_2 produced cling to the sand particles and make them float to the surface.

In wastewater treatments, the flotation method is one of the most important primary procedures. It is also used in the mineralogy industry when crushed rocks are dispersed in water with suitable surfactants (also called *collectors* in the industry) to give stable bubbles on aeration, then hydrophobic minerals are floated to the surface by the attachment of bubbles, while the hydrophilic mineral particles settle to the bottom. The preferential adsorption of the collector molecules on a mineral makes it hydrophobic. Xanthates have been used for the flotation of lead and copper. In these examples, it is the adsorption of xanthane that dominates the flotation.

5 Solid Surface Characteristics
Wetting, Adsorption, and Related Processes

5.1 INTRODUCTION

Experiments have shown that the liquid–solid system exhibits a significant, specific property, the so-called wetting and adsorption–desorption property. Many technical processes related to the characteristics of any solid + liquid system are very important in everyday life. One finds that the wetting (of liquid) characteristics of any solid surface play an important role in all kinds of different systems. For instance, when a fracking water solution is being injected (and when the same solution is being recovered at the borehole), the wetting of the reservoir will be the main characteristics of system. The shale rock is not homogeneous (comprising both inorganic structure and organic (kerogen) structure). The wetting phenomenon will thus be different in different shale reservoirs (Morrow and Mason, 2001; Drummond and Israelachvili, 2004; Kumar et al., 2005; Bolt and Kaldi, 2005; Birdi, 2016; Borysenko et al., 2008; Slatt, 2011).

The next most important step is the process of adsorption–desorption of substances on solid surfaces. These phenomena are the crucial steps for different kinds of systems (such as cleaning processes, oil and gas reservoirs, wastewater treatment, etc.). For example in such systems: wastewater treatment; washing, coatings; adhesion; lubrication; oil recovery; and so on. The surface forces present at the point of liquid and solid contact are found to be related to the respective surface tensions. The liquid–solid or liquid$_1$–solid–liquid$_2$ system is both a contact angle (Young's equation) and capillary phenomenon (Laplace's equation). These two parameters are

$$Cos(\theta) = \frac{(\gamma_S - \gamma_{SL})}{\gamma_L} \tag{5.1}$$

and

$$\Delta P = \frac{(2\gamma_L Cos(\theta))}{Radius} \tag{5.2}$$

In the following, some important phenomena where these parameters are of importance are described.

5.2   OIL AND GAS RECOVERY (CONVENTIONAL RESERVOIRS) AND SURFACE FORCES

In conventional reservoirs, oil is generally found under high temperatures (80°C) and pressures (200 atm [ca. 2 km depth]). In other words, oil in these reservoirs is produced under high-pressure conditions (Appendix I). The pressure needed depends primarily on the porosity of the reservoir rock and the viscosity of the oil, among other factors. This creates a flow of oil through the rocks, which consists of pores of varying size and shape. Roughly, one may compare oil flow with the squeezing of water out of a sponge. One needs to squeeze harder to push water out of a sponge with small pores than out of a sponge with large pores. In the conventional reservoirs, the porosity is much higher than in the case of non-conventional (shale rocks). The degree of oil recovery in all kinds of reservoirs is generally low (much lower than 100%). It is found that around 50% of the oil in place remains unrecovered. This means that all the oil recovered until now leads to some 20%–40% residual oil in the depleted reservoirs. This may be considered as an advantage in the long run, since as the shortage of oil supplies intensifies, we may be forced to develop new technologies to recover the residual oil. In fact, this is the case in most oil wells, which are getting dry (mostly in the North Sea, the USA, etc.) (Appendix I) (Tunio et al., 2011; Birdi, 2009, 2016; Somasundaran, 2015). Physico-chemical methods have been used to improve the degree of recovery. At present, approximately 100 million barrels of oil is used per day. In most oil reservoirs, the primary recovery is based on the natural flow of oil under the gas pressure of the reservoir. In these reservoirs, as this gas pressure drops, then water flooding or other suitable procedures are used. The pressure needed is determined by the capillary pressure of the reservoir and the viscosity of the oil. This procedure still results in 30%–50% of the original resource remaining in the formation. In some cases, one may add such substances as detergents or similar chemicals to enhance the flow of oil through the porous rock structure. The principle is to reduce the capillary pressure (i.e., $\Delta P = \gamma$/curvature) and the contact angle. This process is called *tertiary oil recovery*. The aim is to produce the residual oil, which is trapped in a capillary-like structure in the porous oil-bearing material. The addition of surface-active substances (such as detergents, etc.) reduces the oil–water interfacial tension (IFT; from ca. 30–50 mN/m to less than 10 mN/m). In some cases, a very low IFT has been used. The tertiary process where more complicated chemical additives are designed for a particle reservoir. In all these recovery processes, the IFT between the oil phase and the water phase is the primary parameter. Another important factor is that during the water flooding of an oil reservoir, a large volume of the water phase is found to bypass the oil in the reservoir (Figure 5.1). What this implies is that if one injects water into the reservoir to push the oil, most of the water passes around the oil (the bypass phenomenon) and comes up without being able to push the oil out. This characteristic is a particular challenge at platforms situated at sea, where wastewater treatment is problematic.

The pressure difference to push the oil drop may be larger than to push the water, leading to the so-called bypass phenomenon. In other words, as water flooding is

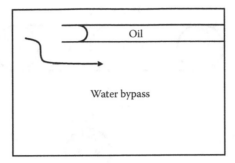

FIGURE 5.1   Water bypass in an oil reservoir.

performed, due to bypass there is less oil produced while more water is pumped back up with oil.

The use of surfactants and other surface-active substances leads to a reduction of $\gamma_{oilwater}$ as described in Figure 5.1. The pressure difference at the trapped oil blob and the surrounding aqueous phase will be

$$\Delta P_{oil\ water} = 2\ IFT\left(\frac{1}{R_1} - \frac{1}{R_2}\right) \tag{5.3}$$

Thus, by decreasing the value of IFT (with the help of surface-active agents) (from 50to <1 mN/m) the pressure needed for oil recovery would be decreased. In the water-flooding process, one uses mixed emulsifiers. Soluble oils are used in various oil well–treatment processes, such as the treatment of water injection wells to improve water injectivity to remove water blockage in producing wells. The same is useful in different cleaning processes on the oil wells. This is known to be effective since water-in-oil microemulsions are found in these mixtures, and with high viscosity. The micellar solution is composed essentially of hydrocarbon, aqueous phase, and surfactant in sufficient quantities to impart micellar solution characteristics to the emulsion. The hydrocarbon is crude oil or gasoline. Surfactants, which are generally used, are of alkyl aryl sulfonate type (more commonly known as petroleum sulfonates).

Another capillary phenomenon to consider is that the pores in the reservoirs are not expected to be perfectly circular (fractal analyses has been reported [Feder, 1988; Birdi, 2003a,b]). In the case of square-shaped pores, one will have to consider bypass in the corners, which are not found in circular pores (Birdi et al., 1988). The magnitude of surface tension, γ, was related to the dimensions of the square capillary and the rise of liquid as follows:

$$\gamma = 1/2\left(d_{liquid}\ g_{gravity}\left(S_{length}\left(C_{constant}\ H_{rise}/2 + C_{constant}\ S_{length}\right)\right)\right) \tag{5.4}$$

where:
   d_{liquid} is the density
   $g_{gravity}$ is the force of gravity

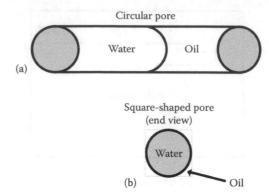

FIGURE 5.2 The bypass phenomenon in: (a) water flooding in an oil reservoir; (b) gas recovery in a shale reservoir (hydraulic fracking).

S_{length} is the side length of the square capillary tubing
H_{rise} is the rise of the liquid in the tubing
$C_{constant}$ (=1.089) is a capillary constant for square shapes

It is interesting to compare this with the circular-shaped capillary data (Equation 2.31). The difference between these two relationships indicates that the magnitude of capillary pressure in complex systems (such as real-world reservoirs) will not be directly related to the idealized circular tubings. The main difference between a circular and a noncircular (e.g., a square shape) pore is that the fluid flow is significantly different (Figure 5.2). In the case of oil–water flooding (Figure 5.2a) the bypass flow is significant (as found in field studies). The same will be the case in hydraulic fracking (Figure 5.2b). The gas will not be able to push back the fracking water to the borehole. This is also observed in actual recovery, where only a fraction of injected water is recovered.

5.2.1 Oil Spills and Clean-Up Process on Oceans

Oil is one of the most transported materials in the world (ca. 100 million barrels per day!). It is transported in tankers across the world between different countries. It is not unexpected, that this has given rise to oil spills in the oceans. Oil platforms situated on the oceans are other sources of oil spills and endanger the natural environment. The worldwide concern with oil spills and their treatment is much dependent on surface chemistry principles (Birdi, 2009, 2016; Somasundaran, 2010, 2015). Oil on sea surfaces will be exposed to various parameters (Figure 5.3):

5.2.2 Different States of Oil Spill on Ocean (or Lakes) Surface

1. Loss of oil by evaporation
2. Loss of oil by sinking to the bottom (as such, or in conjunction with solids)
3. Emulsification (oil–water emulsions)

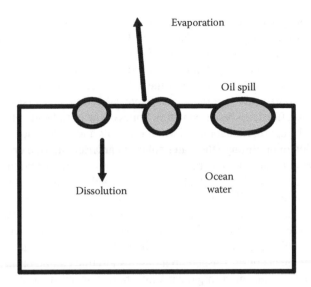

FIGURE 5.3   Oil spill on oceans (different stages of spill treatment).

One of the main treatments is based on surface chemistry applications (use of surface active substances (SAS), etc.). Oil spills are treated by various methods, depending on the region (climate) and the type of oil. Most of the light fluids in oil will evaporate into the air. The oil, which has adsorbed on solid suspension, will sink to the bottom (or float). The remaining oil is generally skimmed off by suitable machines. In some cases, one also uses surfactants to emulsify the oil, and this emulsion sinks to the bottom. However, no two oil spills are the same because of the variations in oil type, location, and weather conditions. Broadly speaking, there are four main methods of response:

1. Leave the oil alone so that it breaks down by natural means. If there is no possibility of the oil polluting coastal regions or marine industries, the best method is to leave it to disperse by natural means. A combination of wind, sun, current, and wave action will rapidly disperse and evaporate most oils. Light oils will disperse more quickly than heavy oils.
2. Contain the spill with booms and collect it from the water surface using skimmer equipment. Spilt oil floats on water and initially forms a slick that is a few millimeters (mm) thick. There are various types of booms that can be used either to surround and isolate a slick, or to block the passage of a slick to vulnerable areas such as the intake of a desalination plant, or fish-farm pens, or other sensitive locations. Boom types vary from inflatable neoprene tubes to solid but buoyant material (most rise up about a meter above the waterline). Some are designed to sit flush on tidal flats, while others are applicable to deeper water and have skirts, which hang down about a meter below the waterline. Skimmers float across the top of the slick contained within the boom and suck or scoop the oil into storage tanks on nearby vessels, or on the shore. However, booms and skimmers are less effective when deployed in high winds and high seas.

3. Use dispersants (SAS) to break up the oil and speed its natural biodegradation. Dispersants act by reducing the IFT at the oil–water interface that inhibits oil and water from mixing. Small droplets of oil are then formed, which helps promote rapid dilution of the oil by water movements. The formation of droplets also increases the oil surface area, thus increasing the exposure to natural evaporation and bacterial action. Dispersants are found to be most effective when used within hours of the initial spill. However, they may not be appropriate for all kinds of oil spills and all locations. Successful dispersion of oil through the water column can affect marine organisms like deep-water corals and sea grass. It can also cause oil to be temporarily accumulated by subtidal seafood. Decisions on whether or not to use dispersants to combat an oil spill must be made in each individual case. The decision will take into account the time since the spill, the weather conditions, the particular environment involved, and the type of oil that has been spilt.

4. Introduce biological agents to the spill to hasten biodegradation. Most of the components of oil washed up along a shoreline can be broken down by bacteria and other microorganisms into harmless substances such as fatty acids and carbon dioxide. This action is called *biodegradation*. Modern oil-spill treatment technology is very advanced and can operate under very divergent conditions. The biggest difference arises from the oil, which may contain varying amounts of heavy components (such as tar, etc.).

5.3   SURFACE CHEMISTRY OF DETERGENCY

The detergent industry is a very large and important area where surface and colloid chemistry principles have been applied extensively. In fact, some detergent manufacturers have been involved in very highly sophisticated research and development work for many decades. Mankind is known to have used soaps for some hundreds of years. The procedure for cleaning fabrics or metal surfaces, and so on, is primarily concerned with removing dirt and so on from the surfaces of clothes (cotton, wool, synthetics, or mixtures of these). It aims to make sure that the dirt does not redeposit after its removal. Dry-cleaning is different, since here one uses organic solvents instead of water. The dirt is adhering to the fabric through different forces (such as Van der Waals and electrostatic). Some components of the dirt are water soluble, and some are water insoluble. The detergents used are designed specifically for these particular processes by the industry and the environment. The composition of the soaps or detergents is mainly based on achieving the following effects:

1. The aqueous solution (water plus added chemicals) should be able to wet the fibbers as completely as possible. This is achieved by lowering the surface tension of the washing water, which thus lowers the contact angle. The low value of surface tension also enables the washing liquid to penetrate the pores (if present), since, from the Laplace equation, the pressure needed would be much lower.

   For example, if the pore size of a fabric (such as any modern type: micro-cotton, Gortex) is 0.3 μm, then it will require a certain pressure

$(=\Delta P = 2\ \gamma/R)$ for water to penetrate the fibers. In the case of water $(\gamma = 72\ \text{mN/m})$, and using a contact angle of $105°$, we obtain:

$$\Delta P = 2\left(72 \times 10^3\right) \text{Cos}\left(105\right)/0.3 \times 10^6$$

$$= 1.4\ \text{bar} \qquad (5.5)$$

2. The detergent then interacts with the dirt/soil to start the process of removal from the fibers and dispersion into the washing water. In order to be able to inhibit the soil once removed to readsorb on the clean fiber, one uses polyphosphates or similar suitable inorganic salts. These salts also increase the pH (to around 10) of the washing water. In some cases, one also uses suitable polymeric anti-redeposition substances (such as carboxymethyl cellulose). The typical compositions of different laundry detergents, shampoos, or dishwashing powders are as follows:

	Laundry Detergent	Shampoo	Dishwashing Powder
Alkyl sulfate	10–20	25	—
Soaps	5	—	—
Nonionics	5–10	—	1–5
Inorganic salts (polyphosphate, silicates)	30–50	—	50
Optical brighteners	<1	—	—

It is worth noting that the aim of detergents in these different formulations is different in each case. In other words, detergents are today tailor-made for each specific application. The detergents in shampoo should give a stable foam in order to increase the cleaning effect. On the other hand, laundry and dishwashing detergents should only give a lower surface tension and almost no foaming (because foaming would reduce the cleaning effect). Hence, in dishwashing formulations, one uses nonionics, which are barely soluble in water and thus produce very little or no foam. These are sometimes of type EOEOEOPOPOPO (ethylene oxide [EO]–propylene oxide [PO]). The propylene group behaves as apolar and the oxide group behaves (through hydrogen bonding) as the polar part. These EOPO types can be tailor-made by combining various ratios of EO:PO in the surfactant molecule. In some cases, even butylene-oxide groups have been used. Furthermore, soil consists mainly of particulate, greasy matter, and so on. The detergents are supposed to keep the soil suspended in the solution and restrict the redeposition. Tests also show that detergents stabilize suspensions of carbon or other solids, such as manganese oxide, in water. This suggests that detergents adsorb on the particles. Furthermore, to these formulations are added redeposition controllers, such as carboxyl-methylcellulose. The detergents are necessary also to remove the greasy part of soil. The adsorption of detergents on soil particles is involved in the detergency process. In the early age of detergent usage during the 1960s, too much sewage treatment showed foaming problems. Later, detergents were used with better degradation properties and better control. For example, straight-chain alkyls were found to be more degradable than branched alkyl chains (Birdi, 1997). This observation has also had consequences for other processes, such as groundwater and fracking processes.

5.4   EVAPORATION RATES OF LIQUID DROPS

In many natural (raindrops, fogs, rivers, waterfalls) and industrial systems (sprays, oil combustion engines, cleaning processes, ink printers, paints, oil spills), one encounters liquid drops—small drops and micro drops—in contact with solid surfaces. The rate of evaporation of liquid from such drops can be of importance in the function of these systems. Extensive investigations on the evaporation of liquid drops (free-hanging drops; drops placed on solid surfaces) have been reported in the current literature (Yu et al., 2004, 2011; Birdi et al., 1989, 1993, 2002; Kim et al., 2007). These drops have been analyzed as a function of

- Liquid (water or organic)
- Solids (plastics; glass)
- Contact angle (θ)
- Height and diameter and volume
- Weight

In these analyses, some assumptions have been made as regards the shape of the drops. The most accurate data arises from using the weight method (Birdi and Vu, 1989). Different analyses showed that the rate of evaporation was linearly dependent on the radius of the drop. The rate of evaporation of a drop is given as a function of the radius (R_{drop}):

$$\text{Rate of evaporation} = 4 \pi R_{drop} \text{ (diffusion constant) (vapor concentration)}$$

On porous solids, the evaporation data showed three different rates of evaporation. The porosity of the solids was found to be related to the third evaporation regime. Furthermore, the apparent contact angle of a water drop on Teflon (i.e., a nonwetting surface) remained constant under evaporation. On the other hand, the apparent contact angle decreased as the water drop evaporated on glass (i.e., a wetting surface) (Figure 5.4).

These analyses were designed to determine the fate of liquid as it evaporates as a function of time. The following different stages would thus be expected:

1. Evaporation of liquid in the free state (as in a normal liquid)
2. Adsorbed liquid state on the solid after state 1

The adsorbed state (2) would thus be dependent on the porous structure of the solid surface. The evaporation studies did indeed show the adsorbed state. From these data, a good correlation to porosity was found (Birdi and Vu, 1989).

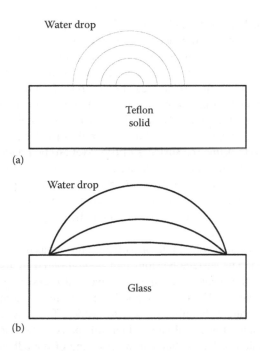

FIGURE 5.4   Profile of a drop of water during evaporation from (a) Teflon surface; (b) glass surface.

5.5   ADHESION (SOLID$_1$–SOLID$_2$) PHENOMENA

In industry and in many other everyday systems, one needs to join two solid surfaces by using glues or adhesives (Kamperman and Synytska, 2012; Bissonnette et al., 2015; Starostina et al., 2014):

- Plastic on metal (car industry, construction industry, etc.)
- Plastic on glass
- Metal to glass
- Wood to wood (furniture, housing, boats)
- Oil- and gas-recovery piping
- Aeroplane wings, windmill wings

Adhesives are based on the solid surface energy characteristics (i.e., polar or nonpolar forces). In the case of the oil- and gas-recovery industry, pipes are coated with different materials (iron–cement, iron–paint, etc.). These coatings are critical to the operation of the piping. The adhesion force is opposite to the energy needed to remove the coating, or the energy needed to break such a contact. For example, in the case of the adhesion of plastic on glass, the greatest adhesion will be obtained if the adhesive fills all the valleys and crevices of each adhered body surface. This will remove any air pockets, which do not contribute to adhesion. The role of the adhesive or glue is to provide mechanical interlocking of the adhesive molecules.

The strength of the bond is dependent upon the quality of this interlocking interface. For achieving optimum bonding, one uses chemical or physical abrading. The abrading process creates many useful properties at the solid surface: enhancement of the mechanical interlocking, creation of a clean surface, formation of a chemically reactive surface, and increase in surface area (a smooth surface has a lower surface area than a rough surface). Diffusion bonding is a form of mechanical interlocking, which occurs at the molecular level in polymers. The science of bonding technology is very extensive. A brief description along with some real examples is given in the following. It is important to prime the surfaces of the layers to be bonded, that is, to cover the surfaces with a dilute solution of the adhesive mixed with an organic solvent to obtain a dried film thickness of 0.0015–0.005 mm and cure separated before applying the adhesive to bond the layers together. Another example is epoxy adhesives. Epoxies are best, but an alloy such as epoxy-phenolic or epoxy-polysulfide may offer improved peeling resistance.

The adhesive is diluted with a suitable solvent until it has the same surface tension as that of the solid. The surface tensions can be compared by using a wetting test, that is, by wetting the surface with the adhesive and measuring the contact angle. A low contact angle ($<90°$) indicates good wetting and an appropriate adhesive. According to mechanical bonding theory, to work effectively, an adhesive needs to fill the valleys and crevices of each adherend (body to be bonded) and displace trapped air. Adhesion is the mechanical interlocking of the adhesive and the adherend together, and the overall strength of the bond is dependent upon the quality of this interlocking interface. Optimum bonding can be achieved by chemical or physical abrading. Abrading the adherend

1. Enhances the mechanical interlocking
2. Creates a clean, corrosion-free surface
3. Forms a chemically reactive surface
4. Increases the bonding surface area

Diffusion bonding is a form of mechanical interlocking, which occurs at the molecular level in polymers. The adsorption mechanism theory suggests that bonding is the process of intermolecular attraction (Vvan der Waals bonding, or permanent dipole, for example) between the adhesive and the adherend at the interface. An important factor in the strength of the bond, according to this theory, is the wetting of the adherend by the adhesive. Wetting is the process by which a liquid spreads onto a solid surface and is controlled by the surface energy of the liquid–solid interface versus the liquid–vapor and the solid–vapor interfaces. In a practical sense, to wet a solid surface, the adhesive should have a lower surface tension than the adherend.

In some systems with *charged surfaces*, the electrostatic forces will have to be considered. Electrostatic forces may also be a factor in the bonding of an adhesive to an adherent. These forces arise from the creation of an electrical double layer of separated charges at the interface and are believed to be a factor in the resistance to separation of the adhesive and the adherend. Adhesives and adherends that contain polar molecules or permanent dipoles are most likely to form electrostatic bonding, according to this theory.

   This theory has been developed to explain the curious behavior of the failure of bonded materials. Upon failure, many adhesive bonds break not at the adhesion interface, but slightly within the adherend or the adhesive, adjacent to the interface. This suggests that a boundary layer of weak material is formed around the interface between the two media. In the following, some mechanisms of adhesive failure are described. The two predominant mechanisms of failure in adhesively bonded joints are adhesive failure or cohesive failure. Adhesive failure is the interfacial failure between the adhesive and one of the adherends. It indicates a weak boundary layer, often caused by improper surface preparation or adhesive choice. Cohesive failure is the internal failure of either the adhesive or, rarely, one of the adherends.

   Ideally, the bond will fail within one of the adherends or the adhesive. This indicates that the maximum strength of the bonded materials is less than the strength of the adhesive strength between them. Usually, the failure of joints is neither completely cohesive nor completely adhesive. It is thus obvious that, for good bonding, the surfaces need to be clean. One needs to remove any dirt, grease, lubricants, water or moisture, and weak surface scales. Solvents are used to clean the soil from solid surfaces. Soil is cleaned from the adherend with an organic solvent that does not affect any of its physical properties. The following different procedures, which have been found useful, are reported:

1. Vapor degreasing is also a good procedure for surface cleaning
2. Solvent wiping, immersion or spraying
3. Ultrasonic vapor degreasing

   The most convenient method is the ultrasonic treatment, with a subsequent solvent rinse of the surface. One also uses other intermediate procedures: abrasive scrubbing, filing, or detergent cleaning.

6 Colloidal Systems

Wastewater Treatment:
Hydraulic Fracking
Technology

6.1 INTRODUCTION

In this chapter, a variety of industrial applications of *colloid chemistry* will be described. Mankind has been aware of the role of finely divided particles for thousands of years. This is particularly noticeable in old structures such as the pyramids or temples (with high structures) or mud (clay, etc.) houses. These structures are typical examples of the role of finely divided particles in stabilizing those structures. The characteristics of finely divided solid particles are dependent on their size and shape. In everyday life, one comes across solid particles of different sizes, ranging from stones on a beach to sand particles, or dust floating around in the air. A special relationship exists between a particle's size (surface area per gram) and its characteristics. Small particles in the size range from 50 Å to 50 μm are called *colloids*. The most simple comparison is sand particles versus dust particles. It is fascinating to observe how dust or other fine particles remain in suspension in the air for long intervals. Once in a while, one observes that a particle experiences a collision-like thrust. In the nineteenth century, it was observed under microscope that small microscopic particles suspended in water made some erratic movements (as if hit by some other neighboring molecules) (Scheludko, 1963; Adamson and Gast, 1997; Chattoraj and Birdi, 1984; Birdi, 1997, 2014, 2016; Somasundaran, 2015). This erratic motion has since been called *Brownian motion*. It arises from the kinetic movement of the surrounding water molecules (Figure 6.1). Thus, colloidal particles remain suspended in solution (for a relatively long time) through Brownian motion, only if gravity forces do not drag the particles to the bottom (or top).

If one throws some sand into the air, one finds that the particles fall to the earth rather quickly. On the other hand, in the case of talcum powder, the particles stay floating in the air for a longer time. This characteristic shows that finely divided solid particles exhibit specific properties as related to their size. The size of particles maybe considered from the following data:

Colloidal dispersions	10 μm–1 nm
Mist/fog	0.1 μm–10 μm
Pollen/bacteria	0.1 μm–10 μm

Oil in smoke/exhaust	1 μm–100 μm
Virus	10 μm–1 nm
Polymers/macromolecules	100 nm–0.1 nm
Micelles	10 nm–0.1 nm
Vesicles	1 μm–1000 μm

The stability of such colloidal systems is a phenomenon that may be roughly compared to a bucket that is stable when standing up but, if tilted beyond a certain angle, topples and comes to rest on its side (Figure 6.2).

The surface forces involved in colloidal particles as found in wastewater are known to be of great significance. Wastewater treatment is an important technology in many everyday phenomena (drinking water, industrial production, etc.). For example, in the case of *fracking*, one uses a suspension of silica in water and other additives. The suspended silica particles are used to assist the stabilization of the fractures in the shale reservoir. The stability of such systems is described by the classical analyses of colloidal systems (Birdi, 1993, 2016). A colloidal suspension may be unstable and exhibit separation of particles within a very short time. Or it may be stable for a very long time, such as over a year. And there will thus be found a *metastable* state, in between these two states. This is an oversimplified statement, but it shows that one should proceed to analyze any colloidal system by following these three states.

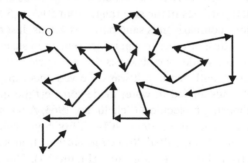

FIGURE 6.1   Brownian motion of a particle.

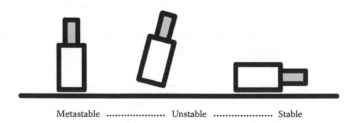

Metastable .... Unstable .... Stable

FIGURE 6.2   Stability criteria of any colloidal system: metastable–unstable–stable states.

As an example, one may consider the wastewater treatment process. Wastewater with colloidal particles is generally a stable suspension. However, by treating it with some definite methods (such as pH control, electrolyte concentration, etc.), one can change the stability of the system. Wastewater treatment technology is one of the most important areas where colloidal surface chemistry is applied. The techniques applied follow a variety of principles, which are strongly related to surface chemistry. The following colloidal forces are needed to describe these systems:

- *Van der Waals forces*: In colloidal systems, the Van der Waals forces play an important role. When any two particles (neutral or charged) come very close to each other, the Van der Waals forces will be strongly dependent on the surrounding medium. For example, in a vacuum, two identical particles will always exhibit attractive forces. On the other hand, if two different particles are present in a medium (in water), then repulsion forces may be present. This can be due to one particle that tends to interact with the medium more strongly than with another particle. One example is silica particles in a water medium with plastics (as in waste water treatment). It is important to understand the conditions under which it is possible that colloidal particles remain suspended. If paint particles (such as titanium dioxide, TiO_2, etc.) aggregate in a container, then the paint is obviously useless. When solid (inorganic) particles are dispersed in an aqueous medium, ions are released into the medium. The ions released from the surface of the solid are of opposite charge. For example, this can be easily shown when glass powder is mixed in water, and one finds that conductivity increases with time. The latter is because minute numbers of ions are released from the glass surface. The presence of the same charge on particles in close proximity gives repulsion, which keeps the particles apart (Figure 6.3).

The positive–positive and negative–negative charged particles will show *repulsion*. On the other hand, the positive–negative charged particles will attract each other. The ion distribution close to the particle surface will also depend on the concentration of any counter-ions or co-ions in the solution. Even glass when dipped in water exchanges ions with its surroundings. Such phenomena can be easily investigated by measuring the change in conductivity of the water.

The force, F_{12}, acting between these opposite charges is given by Coulomb's law (Adamson and Gast, 1997; Scheludko, 1966; Birdi, 2010a, 2016), with charges q_1 and q_2, separated at a distance, R_{12}, in a dielectric medium, D_e:

$$F_{12} = \frac{(q_1 \, q_2)}{(4\pi D_e \, \varepsilon_o R_{122})} \tag{6.1}$$

where ε_o is the electronic charge. The force would be attractive between opposite charges, while repulsive in the case of similar charges. Since the D_e of water is very

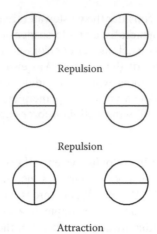

Repulsion

Repulsion

Attraction

FIGURE 6.3   Solid particles (or drops of liquid) with charges (positive–positive or positive–negative or negative–negative).

high (80 units) as compared with the D_e of air (ca 2), one would expect very high dissociation in water, while hardly any dissociation in air or organic liquids (low D_e). As an example, one can estimate the magnitude of F_{12} for Na^+ and Cl^- ions (with a charge of 1.6×10^{-19} C ($=4.8 \times 10^{-10}$ esu)) in water ($D_e = 74.2$ at 37°C), and at a separation distance of (R_{12}) of 1 nm:

$$F_{12} = \left(1.6 \times 10^{-19}\right)\left(1.6 \times 10^{-19}\right) / \left[\left(4\Pi 8.854 \times 10^{-12}\right)\left(10^{-9}\right)\left(74.2\right)\right]$$

$$= -3.1 \times 10^{-21} \, \text{J/molecule} \tag{6.2}$$

where ε_o is 8.854×10^{-12} kg^{-1} m^{-3} s^4 A^2 (J^{-1} C^2 m^{-1}). This gives a value of F_{12} of -3.1×10^{-21} J molecule^{-1} or -1.87 kJ mole^{-1}.

Another very important physical parameter one must consider is the size (and shape) distribution of the colloids. A system consisting of particles of the same size is called *monodisperse*. A system with different sizes is called *polydisperse*. It is also obvious that systems with monodisperse will exhibit different properties than those with polydisperse. In many industrial applications (such as coating on tapes used for recording music, coatings on CDs or DVDs), mono disperse coatings are most preferred. The method used to prepare monodisperse colloids is to achieve a large number of critical nuclei in a short interval of time. This induces all equally sized nuclei to grow simultaneously and thus produce a monodisperse colloidal product.

6.2   COLLOIDS STABILITY THEORY DERJAGUIN–LANDAU–VERWEY–OVERBEEK (DLVO) THEORY: SILICA (PROPPANT) SUSPENSION IN HYDRAULIC FRACKING

The question one needs to understand is under which conditions will a colloidal system remain dispersed for a given length of time (and under other conditions become unstable). How colloidal particles interact with each other is one of the important questions that determines the understanding of the experimental results for phase transitions in such system as found in various industrial processes. The hydraulic fracking solution is a very important example. In the latter systems, additives are used which stabilize the suspensions of silica (5%–10%) in water. One also will need to know under which conditions a given dispersion will become unstable (*coagulation*). For example, one needs to apply coagulation in wastewater treatment such that most of the solid particles in suspension can be removed. Different forces exist when any two particles come close to each other:

ATTRACTIVE FORCES–*REPULSIVE FORCES*

If the attractive forces are larger than the repulsive forces, then the two particles will merge together (unstable system). However, if the repulsion forces are larger than the attractive forces, then the particles will remain separated (stable suspension). It is important to mention here that the medium in which these particles are present thus will, to some degree, contribute also to the stability characteristics. Ionic strength (i.e., concentration of ions) and pH are particularly found to exhibit very specific effects. The different forces of interest are

- Van der Waals
- Electrostatic (if the particles exhibit charges)
- Steric
- Hydration
- Polymer–polymer interactions (if polymers are involved in the system)

In many systems, one may add large molecules (macromolecules; polymers), which when adsorbed on the solid particles, will impart special kinds of stability criteria (Adamson and Gast, 1997; Birdi, 2016). It is well known that neutral molecules, such as alkanes, attract each other mainly through Van der Waals forces. Van der Waals forces arise from the rapidly fluctuating dipoles moment (10^{15} s^{-1}) of a neutral atom, which leads to polarization and consequently to attraction. This is also called the *London potential* between two atoms in a vacuum, and is given as (Scheludko, 1966; Adamson and Gast, 1997; Birdi, 2016)

$$V_{vdw} = -\left(L_{11}/\mathbf{R}^6\right) \tag{6.3}$$

where:

L_{11} is a constant that depends on the polarizability and the energy related to the dispersion frequency

$\mathbf{R}$ is the distance between the two atoms

Since the London interactions with other atoms may be neglected as an approximation, the total interaction for any macroscopic bodies may be estimated by a simple integration.

When two similarly charged colloid particles, under the influence of the electrical double layer (EDL), come close to each other they will begin to interact (Adamson and Gast, 1997; Birdi, 2016). The surface potentials will overlap each other and this will lead to specific consequences. The charged molecules or particles will be under both Van der Waals and electrostatic interaction forces. The Van der Waals forces, which operate at short distances between particles, will give rise to strong attraction forces. The potential of mean force between colloid particles in an electrolyte solution plays a central role in the density, the phase behavior, and the kinetics of agglomeration in colloidal dispersions. This kind of investigation is important in various industries:

- Inorganic materials (ceramics, cements)
- Foods (milk)
- Bio-macromolecular systems (proteins and DNA)

The Derjaguin–Landau–Verwey–Overbeek (DLVO) theory describes that the stability of a colloidal suspension is mainly dependent on the distance between the particles (Adamson and Gast, 1997; Bockris et al., 1981; Birdi, 2003, 2016; Somasundaran, 2015). DLVO theory has been modified in later years and different versions are found in the current literature. Electrostatic forces will give rise to repulsion at large distances (Figure 6.4). This arises from the fact that the electrical charge–charge interactions take place at a large distance of separation. The resultant curve is shown (schematic) in Figure 6.4. The barrier height determines the stability with respect to the quantity k T, the kinetic energy. DLVO theory predicts, in

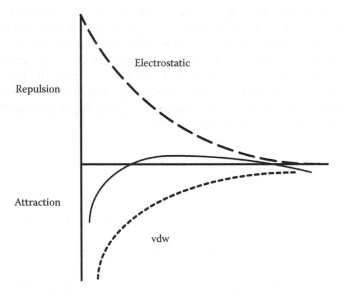

FIGURE 6.4 Variation of repulsion and attraction forces vs. distance between two charged particles (solid line = total force).

most simple terms, that if the repulsion potential (Figure 6.4) exceeds the attraction potential by a value

$$W \gg kT \tag{6.4}$$

then the suspension will be stable. On the other hand, if

$$W \leq kT \tag{6.5}$$

then the suspension will be unstable and it will coagulate. It must be stressed that DLVO theory does not provide a comprehensive analysis. It is basically a very useful tool for such analyses of complicated systems. It is an especially useful guidance theory in any new application or any industrial development.

6.2.1 CHARGED COLLOIDS (ELECTRICAL CHARGE DISTRIBUTION AT INTERFACES)

The interactions between two charged bodies will be dependent on various parameters (e.g., surface charge, electrolyte in the medium, charge distribution) (Figure 6.5). The distribution of ions in an aqueous medium needs to be investigated in such charged colloidal systems. This observation indicates that the presence of charges on surfaces means that a specific surface potential exists. On the other hand, in the case of neutral surfaces, one has only the Van der Waals forces to be considered. This was clearly seen in the case of micelles, where the addition of very small amounts of NaCl to the solution showed

- Large decrease in critical micelle concentration (CMC) in the case of ionic surfactant.
- Almost no effect in nonionic micelles (since in these micelles there are no charges or EDL).

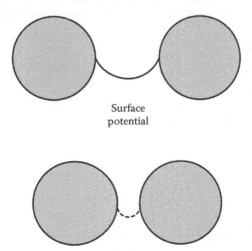

Surface
potential

FIGURE 6.5 Variation of EDL between two charged particles with different ion concentrations (solid line = low; dotted line = high).

In fact, the charged micelle systems clearly indicate the role of surface potential in similar systems. Electrostatic and EDL forces are found to play very important roles in a variety of systems familiar to science and engineering (Bockris et al., 1980; Kortum, 1965; Birdi, 1972, 2010b, 2016; Somasundaran, 2015). It would be useful to consider a specific example to understand these phenomena. Let us take a surface with positive charge, which is suspended in a solution containing positive and negative ions. There will be a definite surface potential, ψ_o, which decreases to a value as the distance of separation increases (Figure 6.6).

It is obvious that the concentration of positive ions will decrease as one approaches the surface of the positively charged surface (charge–charge repulsion). On the other hand, the oppositely charged ions, negative, will be strongly attracted toward the surface. This gives rise to the so-called Boltzmann distribution:

$$n^- = n_o e^{+(z \varepsilon \psi / kT)} \tag{6.6}$$

$$n^+ = n_o e^{-(z \varepsilon \psi / kT)} \tag{6.7}$$

This shows that positive ions are repelled while negative ions are attracted to the positively charged surface. At a reasonably far distance from the particle, $n^+ = n^-$ (as required by the electro-neutrality). The following relationship can be derived for the quantity ψ (r), as a function of distance, r, from the surface as

$$\psi(r) = z \, e / (\mathbf{D} r) \varepsilon - \kappa r \tag{6.8}$$

where κ is related to the ion atmosphere around any ion. In any aqueous solution, when an electrolyte, such as NaCl, is present, it dissociates into positive (Na^+) and negative (Cl^-) ions. Due to the requirement of electro-neutrality (that is, there must be the same number of positive and negative ions), each ion is surrounded by an oppositely charged ion at some distance. Obviously, this distance will decrease with

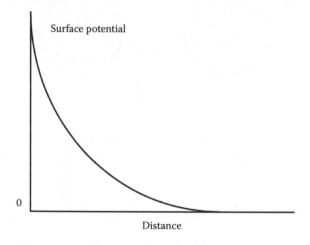

FIGURE 6.6   Variation of surface potential, ψ_o, vs. distance from the solid.

the increasing concentration of the added electrolyte. The expression $1/\kappa$ is called the *Debye length*.

As expected, the Debye-Huckel (D-H) theory tells us that ions tend to cluster around the central ion. A fundamental property of the counter-ion distribution is the thickness of the ion atmosphere. This thickness is determined by the quantity of the Debye length (or Debye radius) $(1/\kappa)$. The magnitude of $1/\kappa$ has a dimension in centimeters, as follows:

$$\kappa = \left(8N^2\right) / \left(1000k_B T\right)^{1/2} I^{1/2} \tag{6.9}$$

where:

$k_B = 1.38 \times 10^{-23}$ J molecule^{-1} K

$e = 4.8 \times 10^{-10}$ esu

Thus, the quantity $k_B T/e = 25.7$ mV at 25°C. As an example, a 1:1 ion (such as: NaCl, KBr, etc.) with a concentration of 0.001 M, gets the value of $1/\kappa$ at 25°C (298 K) (Table 6.1):

$$1/\kappa = \frac{\left(78.3 \times 1.3810^{-16} \times 298\right)}{\left(2 \times 4\Pi \times 6.023 \times 10^{17} \left(4.8 \times 10^{-10}\right)^2\right)^{0.5}}$$

$$= 3 \times 10^{-8} \left(0.001\right)^{1/2} \text{ cm}$$

$$= 9.7 \times 10^{-7} \text{ cm}$$

$$= 97\text{Å} \tag{6.10}$$

The expression in the equation can be rewritten as

$$\psi(r) = \psi_o(r) \exp(-\kappa r) \tag{6.11}$$

TABLE 6.1

Debye Length ($(1/k)$ nm) in Aqueous Solutions (25°C)

Molal	Salt Concentration			
	1:1	1:2	2:2	1:3
0.0001	30.4	17.6	15.2	12.4
0.001	9.6	5.55	4.81	3.9
0.01	3.04	1.76	1.52	1.2
0.1	0.96	0.55	0.48	0.4

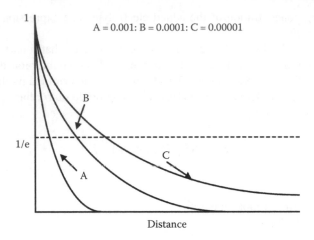

FIGURE 6.7   Variation (decrease) in electrostatic potential with distance of separation as a function of electrolyte concentration (ionic strength).

which shows the change in ψ (r) with the distance between particles (r), and at a distance of $1/\kappa$, the potential has dropped to ψ_0. This is accepted to correspond with the thickness of the double layer. This is the important analyses, since the particle–particle interaction is dependent on the change in $\psi(r)$. The decrease in $\psi(r)$ at the Debye length is different for different ionic strength, Figure 6.7.

The data in Table 6.1 show values of D-H radius in various salt concentrations. The magnitude of $1/\kappa$ decreases with I and with the number of charges on the added salt. This means that the thickness of the ion atmosphere around a reference ion will be much compressed with the increasing value of I and z_{ion}. A trivalent ion such as Al^{3+} will compress the double layer to a greater extent in comparison with a monovalent ion such as Na^+. Furthermore, inorganic ions can interact with a charged surface in one of two distinct ways:

1. Nonspecific ion adsorption where these ions have no effect on the iso-electric point
2. Specific ion adsorption, which gives rise to a change in the value of the isoelectric point

Under those conditions, where the magnitude of $1/\kappa$ is very small (for example in high electrolyte solution), then one can write:

$$\psi = \psi_0 \exp - (\kappa x) \tag{6.12}$$

where x is the distance from the charged colloid.

The value of ψ_0 is found to be 100 mV (in the case of monovalent ions) (=4 **kB** **T**/z e). Experimental data and theory shows that the variation of ψ is dependent on the concentration and the charge of the ions. These relationships thus show

- That the surface potential drops to zero at a faster rate if the ion concentration (C) increases.
- That the surface potential drops faster if the value of z goes from 1 to 2 (or higher)

These theories are found to agree with experimental data. In many industrial applications, the surface charge on particles is found to be of primary importance.

6.2.2 ELECTROKINETIC PROCESSES OF CHARGED PARTICLES IN LIQUIDS

In the following, let us consider what happens if the charged particle or surface is under *dynamic motion* (in water) of some kind. There are different systems under which the electrokinetic phenomena are investigated. These systems are (Kortum, 1965; Bockris et al., 1980; Adamson and Gast, 1997; Birdi, 2016)

1. Electrophoresis: This system refers to the movement of the colloidal particle under an applied electric field.
2. Electro-osmosis: This system is where a fluid passes next to a charged material. This is actually the complement of electrophoresis. The pressure needed to make the fluid flow is called the *electro-osmotic pressure.*
3. Streaming potent: If fluid is made to flow past a charged surface, then an electric field is created, which is called *streaming potential.* This system is thus the opposite of the electro-osmosis.
4. Sedimentation potential: A potential is created when charged particles settle out of a suspension. This gives rise to sedimentation potential, which is the opposite of the streaming potential.

The reason for investigating the electrokinetic properties of a system is to determine the quantity known as the zeta potential.

Electrophoresis is the movement of an electrically charged substance under the influence of an electric field. This movement may be related to the fundamental electrical properties of the body under study and the ambient electrical conditions by Equation 6.13. F is the force, q is the charge carried by the body, and E is the electric field:

$$Fe = qE \qquad (6.13)$$

The resulting electrophoretic migration is countered by forces of friction such that the rate of migration is constant in a constant and homogeneous electric field:

$$F_f = vf_r \qquad (6.14)$$

where v is the velocity and fr is the frictional coefficient.

$$QE = vf_r \qquad (6.15)$$

The electrophoretic mobility μ is defined as followed:

$$\mu = \frac{v}{E} = \frac{q}{f_r} \tag{6.16}$$

This expression applied only to ions at a concentration approaching 0 and in a nonconductive solvent. Polyionic molecules are surrounded by a cloud of counterions that alter the effective electric field applied on the ions to be separated. This renders the previous expression a poor approximation of what really happens in an electrophoretic apparatus. The mobility depends on both the particle properties (e.g., surface charge density and size) and the solution properties (e.g., ionic strength, electric permittivity, and pH). For high ionic strengths, an approximate expression for the *electrophoretic mobility*, μ_e, is given by the following equation (Adamson and Gast, 1997; Birdi, 2016):

$$\mu_e = \varepsilon \varepsilon_o \, \eta / \zeta \tag{6.17}$$

where:
     ε is the dielectric constant of the liquid
     ε_o is the permittivity of free space
     η is the viscosity of the liquid
     ζ is the zeta potential (i.e., surface potential) of the particle

6.3   STABILITY OF LYOPHOBIC SUSPENSIONS

Particles in all kinds of suspensions or dispersions interact with two different kinds of forces (e.g., attractive forces and repulsive forces). One observes that lyophobic suspensions (sols) must exhibit a maximum in repulsion energy in order to have a stable system. The total interaction energy, V(h), is expressed as (Scheludko, 1966; Bockris et al., 1980; Attard, 1996; Adamson and Gast, 1997; Hsu and Kuo, 1997; Chattoraj and Birdi, 1984; Birdi, 2002, 2016; Somasundaran, 2015)

$$V(h) = V_{el} + V_{vdw} \tag{6.18}$$

where:
   V_{el} is electrostatic repulsion components
   V_{vdw} is Van der Waals attraction components

Dependence of the interaction energy $\mathbf{V}$(h) on the distance $\mathbf{h}$ between particles, has been ascribed to coagulation rates as follows:

1. During slow coagulation
2. When fast coagulation sets in

The dependence of energy V(h) on h is given as

$$V(h) = \left[\left(64 \, C \, RT\psi^2 \right) / k \; \exp(-kh) - H / \left(2h^2 \right) \right] \tag{6.19}$$

For a certain ratio of constants, it has the shape shown in Figure 6.8. For large values of $\mathbf{h}$, V(h) is negative (attraction), following the energy of attraction V_{vdw}, which decreases more slowly with increasing distance ($\sim 1/\mathbf{h}^2$). At short distances (small h), the positive component V_{el} (repulsion), which increases exponentially with decreasing $\mathbf{h}(\exp(-k\,\mathbf{h})$, can overcompensate V_{vdw} and reverse the sign of both dV(h)/dh and V(h) in the direction of repulsion. On further reduction of the gap (very small h), V_{vdw} will again predominate, since

$$V_{el} \rightarrow 64\ C\ RT\psi^2/k,\ \text{as}\ h \rightarrow 0, \qquad (6.20)$$

whereas the magnitude of the quantity V_{vdw} increases indefinitely when $h > 0$. There is thus a repulsion maximum in the function V(h), which can be estimated from the condition dV(h)/dh$=0$. The choice of solution (maximum or minimum) does not present any difficulty since V(h) is positive for the maximum. From these considerations, it is found that when the electrolyte concentration is increased, the magnitude of k in the exponent of V_{el} also increases (compression of diffuse layers), so that the maximum caused by it becomes lower. At a certain value of C, the curve V(h) will become similar to curve (B) in Figure 6.8, with $V(h)_{max}=0$.

Experiments have shown that the coagulation rate will become fast starting from this concentration. This is called the *critical concentration*, C_{cc}. In other words, C_{cc} can be estimated from simultaneous solution of the following:

$$dV(h)/dh = 0 \text{ and } V(h) = 0. \qquad (6.21)$$

One can write the following:

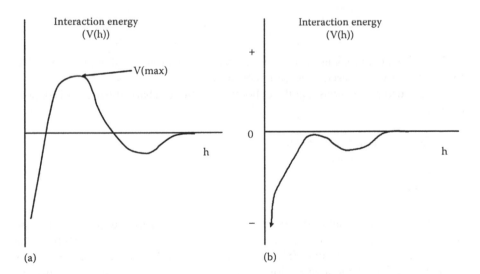

(a)   (b)

FIGURE 6.8 Plot of interaction energy, V(h), vs. distance, h, between particles: (a) slow coagulation; (b) fast coagulation.

$$\frac{dV(h)}{dh} = -\left(64\,C_{cc}RT\psi^2\right)/k\ \exp\left(-k_{cr}h_{cr}\right) + K/\left(h_{cr}^3\right) = 0 \qquad (6.22)$$

and:

$$V(h) = \left[\left(64\,C_{cc}RT\psi^2\right)/k_{cr}\ \exp\left(-k_{cr}h_{cr}\right) - K/\left(2\,h_{cr}2\right) = 0\right] \qquad (6.23)$$

After expanding these expressions, as related to h and C, this becomes (the Schulze–Hardy rule) (for suspensions of charged particles in water):

$$C_{cc} = 8.7\times10^{-39}/Z^6A^2$$

$$C_{cc}Z^6 = k_{constant} \qquad (6.24)$$

where the $k_{constant}$ includes (Hamaker constant $=4.2\times10^{-19}$ J) and other constants, except Z and critical concentrations of added ion (C_{cc}). The magnitude of Z^6 for different ions is

$$Z^6 = 1:\left(2^6\right)0.016:\left(3^6\right)0.0014:\left(4^6\right)0.000244 \qquad (6.25)$$

The experimental data for various colloidal systems, such as As_2S_3 or Au (sol), have given values, which agree with the relation in Equation 6.25:

Valency of Counter-Ion	Magnitude of Ccc	Ratio
Monovalent	50	1:1
Divalent	0.8	$1:2^6$
Trivalent	0.08	$1:3^6$

These results are some of the most convincing data which show that the EDL theory of colloids is based on a useful correlation to real systems.

It thus becomes obvious that the colloidal stability of charged particles is dependent on

1. Concentration of electrolyte
2. Charge on the ions
3. Size and shape of colloids
4. Viscosity

The C_{cc} is thus found to depend on the type of electrolyte used as well as on the valency of the counter-ion. It is seen that divalent ions are 60 ($=2^6$) times as effective as monovalent ions. Trivalent ions are several hundred ($729=3^6$) times more effective than monovalent ions. Experiments have shown that this correlation to C_{cc} is almost exact and the assumptions made in Equations 6.23 and 6.24 are thus useful for understanding such systems. However, ions that specifically adsorb (such as surfactants) will exhibit different behavior, as also found from experiments. For example, based on these observations, in the composition of washing powder, multivalent phosphates are used, for instance, to keep the charged dirt particles from attaching to the fabrics after having been removed. Another example is wastewater treatment, where for coagulation purposes one uses multivalent ions.

Streaming potential: The interface of a mineral (rock) in contact with the aqueous phase exhibits a surface charge. The currently accepted model of this interface is the EDL model of Stern (Adamson and Gast, 1997; Birdi, 2014, 2016). Chemical reactions take place between the minerals and the electrolytes in the aqueous phase, which results in a net-charge on the mineral. Water and electrolytes bound to the rock surface constitute the Stern (or Helmholtz) layer. In this region, the ions are tightly bound to the mineral, while away from this layer (the so-called diffuse layer), the ions are free to move about.

Since the distribution of ions (positive and negative) is even in the diffuse region, there is no net-charge. On the other hand, in the Stern layer there will be asymmetric charge distribution. As one will measure from zeta potential data, the mineral exhibits a net-charge.

6.3.1   KINETICS oF COAGULATION oF COLLOIDS

Colloidal solutions are characterized by the degree of stability or instability. This is related to the fact that one needs to understand both kinds of properties in everyday phenomena. The kinetics of coagulation is studied by using different methods. The number of particles, N_p, at a given time is dependent on the diffusion-controlled process. The rate is given by

$$-\frac{dN_p}{dt} = 8\pi \mathbf{DR} N_p^2 \qquad (6.26)$$

where:
   $\mathbf{D}$ is the diffusion coefficient
   $\mathbf{R}$ is the radius of the particle

The rate can be rewritten as

$$-\frac{dN_p}{dt} = \frac{4k_B T}{3 v\, N_p^2} = k_o N_p^2 \qquad (6.27)$$

where:
   $\mathbf{D} = k_B T/6\pi\, v\, \mathbf{R}$ (Einstein equation)
   k_o is the diffusion-controlled constant

In real systems, one is interested in stable colloidal systems (as in paints, creams) while in other cases, one is interested in unstable systems (as in wastewater treatment).

It is thus seen that from DLVO considerations, the degree of colloidal stability will be dependent on the following factors:

1. Size of particles (larger particles will be less stable)
2. Magnitude of surface potential
3. Hamaker constant (H)
4. Ionic strength
5. Temperature

The attraction force between two particles is proportional to the distance of separation and a Hamaker constant (specific to the system). The magnitude of H is of the order of 10^{-12} erg (Adamson and Gast, 1997; Birdi, 2002, 2016; Somasundaran, 2015). The DLVO theory is thus found useful to predict and estimate the colloidal stability behavior. Of course, in such systems with many variables, this simplified theory is expected to fit all kinds of systems. In the past decade, much development has taken place as regards measuring the forces involved in these colloidal systems. In one method, the procedure used is to measure the force present between two solid surfaces at very low distances (less than a micrometer) (Birdi, 2003). The system can operate under water, and thus the effect of additives has been investigated. These data have provided verification of many aspects of DLVO theory.

Recently, an atomic force microscope (AFM) has been used to directly measure these colloidal forces (Birdi, 2003) (see Chapter 7). In AFM, two particles are brought closer (nm distance) and the force (nanonewton) is measured. In fact, commercially available apparatuses are designed to perform such analyses. The measurements can be carried out both in air and under fluids, and under various experimental conditions (such as added electrolytes, pH, etc.) (Birdi, 2003).

6.3.2 Flocculation and Coagulation of Colloidal Suspension

In everyday life, one finds a variety of systems where solid particles are suspended in water (such as wastewater plants, fracking water, etc.). It is known from common experience that a colloidal dispersion with smaller particles is more stable than one with larger particles. The phenomenon of smaller particles forming aggregates of a larger size is called *flocculation* or *coagulation*. For example, to remove insoluble and colloidal metal precipitates, one uses flocculation. This is generally achieved by reducing the surface charges, which gives rise to weaker charge—charge repulsion forces. As soon as the attraction forces (vdW) become larger than the electrostatic forces, coagulation takes place. Coagulation is initiated by particle charge neutralization (by changing pH or other methods [such as charged polyelectrolytes]), which leads to aggregation of particles to form larger particles. This means that:

Initial state: charge—charge (repulsion)
Final state: neutral—neutral (attraction) (coagulation)

Coagulation can also occur by adding suitable substances (coagulants) particular for a given system. The latter reduce the effective radius of the colloid particle and lead to coagulation.

Flocculation is a secondary process after coagulation, and leads to very large particle (floccs) formation. Experiments show that coagulation takes place when the zeta potential is around ±0.5 mV. Coagulants such as iron and aluminum inorganic salts are effective in most cases. For example, in wastewater treatment plants, the zeta potential is used to determine the coagulation and flocculation phenomena. In general, most of the solid material in wastewater is negatively charged.

6.4 WASTEWATER TREATMENT AND CONTROL (ZETA POTENTIAL)

Wastewater contains different kinds of pollutants (dissolved substances, suspended particles). In most shale gas hydraulic fracture processes, wastewater treatment is needed (Slatt, 2011; Chapman et al., 2012). In a typical horizontal well gas shale, one may use 2×10^4 m^3 of water containing additives (with concentrations less than 1%). However, about less than half of this volume of water is recovered as flowback, which is partially mixed with brine water, as present in the reservoir. The wastewater is treated in suitable plants before the processed water is released into the surroundings (or reused in most cases). The produced water contains a high amount of total dissolved solids (TDS) (Shih et al., 2015). The wastewater contains pollutants that are both soluble and insoluble substances, which need to be removed. The substances that are found in wastewater (solutes) are in either molecular state (such as benzene, oils, etc.) or ionic form (such as Na^+, Cl^-, Mg^{++}, K^+, Fe^{++}, etc.). The concentrations of pollutants are generally given in various units:

Weight/volume	mg L^{-1}; kg m^{-3}
Weight/weight	mg kg^{-1}; parts per million (ppm); parts per billion (ppb)
Molarity	moles L^{-1}
Normality	Equivalents L^{-1}

The specific unit used depends on the amounts present. The unit used for trace amounts, such as benzene is given in parts per million (ppm) or parts per billion (ppb). The hardness of drinking water (mostly Na; Ca-Mg) concentration is given as milligrams per liter (mg L^{-1}). The typical values found are in the range of less than 10 mg L^{-1} (soft water) or over 20 mg L^{-1} (hard water).

The presence of a net-charge at the particle surface gives rise to an asymmetric distribution of ions in the surrounding region. This means that the concentration of counter-ions close to the surface is higher than the ions with the same charge as the particle. Thus, an EDL is measured around such particles placed in water.

The solids can be removed by filtration and precipitation methods. The precipitation (of charged particles) is controlled by making the particles flocculate by controlling the pH and ionic strength. The latter gives rise to a decrease in charge–charge repulsion and thus can lead to precipitation and the removal of finely divided suspended solids. It is thus found that the most important factor that effects the *zeta potential* is pH. Therefore, all zeta potential data must mention their pH. Imagine a particle in suspension with a negative zeta potential. If more alkali is added to this suspension, then the particle will exhibit an increase in negative charge. On the other hand, if acid is added to the colloidal suspension, then the particle will acquire increasing positive charge. During this process, the particle will undergo a change from negative charge to *zero* charge (where the number of positive charge is equal to negative charge [*point-of-zero-charge*: PZC]). In other words, one can control the magnitude and sign of the surface charge by a potential determining ion.

The stability is dependent on the magnitude of the electrostatic potential at the surface of the colloid, ψ_0. The magnitude of ψ_0 is estimated by using the microelectrophoresis method. When an electric field is applied across an electrolyte, charged particles suspended in the electrolyte are attracted toward the electrode of opposite charge. Viscous forces acting on the particles tend to oppose this movement. When equilibrium is reached between these two opposing forces, the particles move with constant velocity. In this technique, the moving (or rather, the speed) of a particle is observed under a microscope when subjected to an electric field. The field is related to the applied voltage, **V**, divided by the distance between the electrodes (in cm). The velocity is dependent on the strength of the electric field or the voltage gradient, the dielectric constant of the medium, the viscosity, and the zeta potential. Commercially available electrophoresis instruments are used where the quartz cells designed for any specific system are available. The magnitude of the zeta potential, ζ, is obtained:

$$\zeta = \mu\eta/\varepsilon_o\mathbf{D} \qquad (6.29)$$

where:

- η is the viscosity of the solution
- ε_o is the permittivity of the free space
- $\mathbf{D}$ is the dielectric constant

The velocity of a particle in a unit electric field is related to its electric mobility. In another application, the magnitude of the zeta potential is measured as a function of added counter-ions. The variation in the zeta potential is found to be related to the stability of the colloidal suspension. The results of a gold colloidal suspension (gold [Au] sol) are reported as follows as a function of counter-ion (Al) concentration:

Counter-Ion Al + 3	Velocity	Stability of Colloidal Au
0	3(−)	Very high
20×10^{-6} mole	2.(−)	Flocculate (4 h)
30×10^{-6} mole	0.(zero)	Flocculates fast
40×10^{-6} mole	0.2(+)	Flocculate (4 h)
70×10^{-6} mole	1(+)	Flocculates slowly

These data show that the colloidal particles charge changes from negative to zero (when the particles do not show any movement) to positive, at high counter-ion concentration. This is a very general picture. Therefore, in wastewater treatment plants, one adds counter-ions until the movement of particles is almost zero and thus one can achieve fast flocculation of pollutant particles. The variation of ζ of silica particles has been investigated as a function of pH. The dissociation of the surface groups $-Si(OH)$ is involved in these characteristics. Under these operations, one constantly monitors the zeta potential by using a suitable instrument.

7 Foams and Bubbles
Formation, Stability and Application

7.1 INTRODUCTION

The formation and structure of thin liquid films (TLF), such as in foams or bubbles, is the most fascinating phenomenon that mankind has studied over many decades. It could be claimed that the TLF structure is the closest one comes to observing molecular structures with the naked eye. TLF is thus the thinnest object one can see without the aid of any kind of microscope. One of the best-known liquid thin film structures is the soap bubble, or bubbles formed from detergent solutions (as in dishwashing solutions). Everyone enjoys watching the formation of soap bubbles and viewing the rainbow colors. Bubble formation and stability may not seem of much consequence, but, in fact, in everyday life bubbles play an important role (e.g., from determining lung function to making beer and champagne!). It is a common observation that ordinary water when shaken does not form any bubbles at the surface. On the other hand, all soap and detergent solutions, and many liquids (shampoo, washing-up liquid, beer, champagne, seawater) on shaking may form very extensive bubbles. In this chapter, the formation and stability of bubbles will be described. Furthermore, even though one cannot see or observe the surface layer of a liquid directly, the TLFs allow one to make some observations that provide much useful information (Somasundaran et al., 1981; Ivanov, 1988; Wilson, 1989; Rubinstein and Bankoff, 2001; Birdi, 2003, 2016).

7.2 BUBBLES AND FOAMS

If pure water is shaken, no bubbles are observed at the surface. Pure organic fluids exhibit no bubble formation on shaking. That means that as an air bubble rises to the surface of the liquid, it merely exits into the air. However, if an aqueous detergent (surface-active substance [SAS]) solution is shaken or an air bubble is created under the surface, then a bubble is formed (Figure 7.1).

This process can be described as follows:

- Air bubble inside liquid phase: At surface, bubble detaches and moves up under gravity.
- The detergent molecule forms a bilayer in the bubble film. The water in between is the same as the bulk solution. This may be depicted as follows:

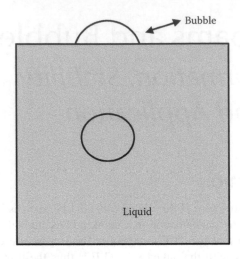

FIGURE 7.1 Formation of a bubble.

- Surface layer of detergent
- Bubble with air and a layer of detergent
- Bubble at the surface forms double layer of detergent with some water in between (TLF) (varying from 10 to 100 μm)

In fact, this test (i.e., the formation of bubbles if a water solution is shaken) is very sensitive and may be used to determine the presence of very minute (few ppm) contents of SASs. A bubble is composed of a TLF with two surfaces, each with a polar end pointed inward and the hydrocarbon chains pointing outward. The water inside the films will move away (due to gravity) giving rise to the thinning of the film. Since the thickness approaches the dimensions of the light wavelength, one observes varying interference colors. The reflected ray will interfere with the incident wavelength. The consequence of this will be that, depending on the thickness of the film, one will observe colors. Especially, when the thickness of the film is approximately the same as the wavelength of the light (i.e., 400–1000 Å). The black film is observed when the thickness is the same as the wavelength of the light (approximately 500–700 Å). Thus, this provides the closest visual observation of two-molecule-thick film by the naked eye.

7.2.1 Application of Foams and Bubbles in Technology

Some of the most important roles for bubbles are found in the food industry (ice cream, champagne, and beer). The stability and size of the bubbles determine the taste and appearance of the product. In this industry, much research has been conducted on the determination of the factors that control bubble formation and stability. During the making of ice creams, air bubbles are trapped in the frozen material. Since the concentration of SAS is very high at the surface, the bubble can be used to remove the latter from the solution. This observation gives rise to a large area of applications where foam-bubble formation is used.

7.3   FOAMS (THIN LIQUID FILMS)

Ordinary foams from detergent solutions are thick initially (measured in micrometers), and as fluid flows away due to gravity or capillary forces or surface evaporation, the film becomes thinner (a few hundred Å). The foam consists of

- Air on one side
- Outer monolayer of detergent molecule
- Some amount of water
- Inner monolayer of detergent molecule
- Air on outer

This can be depicted (schematic) as follows:

DETERGENT*WATER*DETERGENT
DETERGENT*WATER*DETERGENT
DETERGENT*WATER*DETERGENT
DETERGENT*WATER*DETERGENT

(thickness varies from 100 μm to a few hundred Å)

The orientation of a detergent molecule in TLF is such that the polar group (OO) is pointing toward the water phase and the apolar alkyl part (CCCCCCCCC) is pointing toward the air.

- AirCCCCCCCCCCOO*WATER*OOCCCCCCCCCCAir
- AirCCCCCCCCCCOO*WATER*OOCCCCCCCCCCAir
- AirCCCCCCCCCCOO*WATER*OOCCCCCCCCCCAir
- AirCCCCCCCCCCOO*WATER*OOCCCCCCCCCCAir
- AirCCCCCCCCCCOO*WATER*OOCCCCCCCCCCAir
- AirCCCCCCCCCCOO*WATER*OOCCCCCCCCCCAir

The thickness of the water phase can vary from over 100 μm to less than 100 nm. Foams are thermodynamically unstable, since there is a decrease in total free energy when they collapse. As the thickness reduces to around the wavelength of light (nm), one starts to observe rainbow colors (arising from interference), and the TLF at even smaller thickness (50 Å or 5 nm). However, certain kinds of foams are known to persist for very long periods, and many attempts have been made to explain their metastability. The TLF may be regarded as a kind of condenser. The repulsion between the two surfactant layers (Figure 7.2), will be determined by the electric double layer (EDL). The effect of added ions to the solution is to make the EDL contract and this leads to thin films.

It looks black-gray and the thickness is around 50 Å (5 nm), which is almost the size of the bilayer structure of the detergent (i.e., twice the length [ca. 25 Å] of a typical detergent molecule). It is remarkable that one can see a two-molecule-thin structure. The rainbow colors are observed since the light is reflected by the varying thickness of the TLF of the bubble. It may not be obvious at first sight, but in the

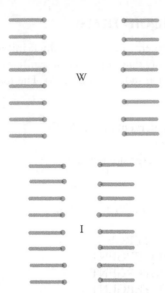

FIGURE 7.2 Thickness of foam films: in water (W) and with added electrolyte (I).

beer industry, foaming is one of the most important characteristics. A beer bottle is produced under high pressure of CO_2 gas. As soon as one opens a beer bottle, the pressure drops and the gas, CO_2, is released, which gives rise to foaming. Usually, the foam stays inside the bottle. The foaming is caused due to the presence of different amphiphilic molecules (fatty acids, lipids, proteins). This foam is very rich as the liquid-film is very thick and contains lots of aqueous phase (such foams are called *kugelschaum*). The foam fills the empty space in the bottle, and, under normal conditions, it barely spills out. However, under certain abnormal conditions, the foam is highly stable and starts to pour out of the bottle and is considered undesirable (Birdi, 1989; Clark et al., 1994). As regards foam stability, it was recognized that the surface tension under film deformation must always change in such a way as to resist the deforming forces. Thus, tension in the film where expansion takes place will increase, while it will decrease in the part where contraction takes place. There is, therefore a force tending to restore the original condition. The film elastic, the term *elasticity* has been defined as:

$$E_{film} = 2A(d\gamma/dA) \tag{7.1}$$

where:

E_{film} is the elasticity of the film

A is the area of the film

γ is the surface tension of the surface deformed

Bubble formation in champagne is another important area of application. The size and number of bubbles affects the taste; bubble stability also has a significant impact on the taste and appearance. The stability of any foam film is related to the kinetics of the thinning of the TLF. As the thickness reaches a critical value, so the stability becomes critical. It was recognized at a very early stage in the research that the unstable state will conform when the film diverges from bulk system properties (Scheludko, 1966; Birdi, 2016). This thickness was estimated to be in the range of 50–150 Å, and is called the *black film state*. In this state, the random motion of the molecules may easily give rise to a rupture of the TLF. The following factors determine the drainage of the fluid in these films:

- Flow of liquid under gravity: It was found that, assuming the fluid in the film has the same viscosity and density, then the mean velocity will not exceed 1000 D_{film}^2, where latter is the distance between the lamella:

D (mm)	Flow (mm/s)
0.01	0.1
0.001	0.001

- Suction flow due to curvature: This effect can be much greater than the gravity effect.
- Evaporation loss: The evaporation process will depend on the surroundings, and will be almost negligible in a closed container.

7.3.1 FOAM STABILITY

As we know, if one blows air bubbles in pure water, no foam is formed. On the other hand, if a detergent or protein (amphiphile) is present in the system, adsorbed surfactant molecules at the interface give rise to foam or soap bubble formation. Foam can be characterized as a coarse dispersion of a gas in a liquid, where gas is the major phase volume. The foam, or the lamina of liquid, will tend to contract due to its surface tension, and a low surface tension would thus be expected to be a necessary property for good foam forming. Furthermore, in order to be able to stabilize the lamina, it should be able to maintain slight differences of tension in its different regions. It is also therefore clear that a pure liquid, which has constant surface tension, cannot meet this requirement. The stability of such foams or bubbles has been related to the monomolecular film structures and stability. For instance, foam stability has been shown to be related to the surface elasticity or surface viscosity, η_s, besides other interfacial forces (Chattoraj and Birdi, 1984; Birdi, 2016). Foam destabilization has also been found to be related to the packing and orientation of mixed films, which can be determined from monolayer studies. It is also worth mentioning that foam formation from monolayers of amphiphiles constitutes the most fundamental process in everyday life. The other assemblies, such as vesicles and bilayer membrane (BLM) are somewhat more complicated systems, which are also found to be in equilibrium with monolayers (Birdi, 2016).

Although the surface potential, ψ, the electrical potential due to the charge on the monolayers, will clearly affect the actual pressure required to thin the lamella to any given thickness, one may assume for the purpose of a simple illustration that $1/k$, the mean Debye–Hückel thickness of the ionic double layer, will influence the ultimate thickness when the liquid film is under a relatively low pressure. Let us also assume that each ionic atmosphere extends only to a distance $3/k$ into the liquid when the film is under a relatively low excess pressure from the gas in the bubbles; this value corresponds to a repulsion potential of only a few millivolts. This assumption gives the following relationship (1 atm pressure) (Wilson, 1989; Birdi, 2016):

$$h_{film} = 6/k + 2 \left(\text{monolayer thickness}\right) \qquad (7.2)$$

For charged monolayers adsorbed from 10^{-3} n sodium oleate, the final total thickness, h_{film}, of the aqueous layer should thus be of the order 600 Å (i.e., $6/k$ Å). To this value, one needs to add 60 Å ($=10^{-10}$ m) for the two films of oriented soap molecules, giving a total of 660 Å. The experimental value is 700 Å. The thickness decreases on the addition of electrolytes, as also suggested by Equation 7.2.

For instance, the value of h_{film} is 120 Å in the case of 0.1 M-NaCl. The addition of a small amount of certain nonionic surface-active agents (e.g., n-lauryl alcohol, n-decyl glycerol ether, laurylethanolamide, laurylsufanoylamide) to anionic detergent solutions has been found to stabilize the foam. It was suggested that the mode of packing is analogous to the palisade layers of the micelles and the surface layers of the foam lamellae. The thickness of foams from two different detergent solutions has been reported (Scheludko, 1966):

Detergent Solution	Black Film Thickness	Molecular Dimension
Sodium oleate	40 Å	36 Å
Octyl-phenoxy-ethanol	65 Å	67 Å

These data convincingly show that the thickness of the black film is the bilayer structure. Measurements have been carried out on the excess tensions, equilibrium thicknesses, and compositions of aqueous foam films stabilized by either n-decyl methyl sulfoxide or n-decyl trimethyl ammonium-decyl sulfate and containing inorganic electrolytes. It has been reported (Scheludko, 1966; Birdi, 2003, 2007, 2016) that the stability of a liquid film must be greatest if the surface pressure strongly resists deforming forces. In the event of the area of the film being extended by a shock (or vibration), then the change in surface pressure, Π, is given as

$$\Pi = -\left(d\Pi/dA\right)\left(A2 - A1\right) \qquad (7.3)$$

where A1 and A2 are, respectively, the available areas per molecule of the foam stabilizing agent in the original and in the extended parts of the surface. This can be written as

$$\Pi = -A1(d\Pi/dA)(A2/A1-1) \qquad (7.4)$$

$$= -A1(d\Pi/dA)(j-1) \qquad (7.5)$$

$$= Cs^{-1}(j-1) \qquad (7.6)$$

where j is the area extension factor. The term $(-A\{d\Pi/dA\})$ is the surface compressibility modulus of the monolayer. For a large restoring pressure Π, this modulus should be large. In the extended region, the local reduction of the surface pressure to (Π) results in a spreading of molecules from the adjacent parts of the monolayer to the extended region. The tension of TLF can be measured by applying pressure and measuring the radius of curvature. Then, using Laplace's equation, one can estimate the tension.

It has been shown (Friberg, 1976; Birdi, 2002, 2007) that a correlation exists between foam stability and the elasticity (**E**) of the film, that is, the monolayer. In order for **E** to be large, surface excess must be large. Maximum foam stability has been reported in systems with fatty acid and alcohol concentrations well below the minimum in γ. Similar conclusions have been observed with n-$Cl_2H_{25}SO_4Na$ (SDS)+$Cl_2H_{25}OH$ systems, which give a minimum in γ versus concentration with maximum foam at the minimum point (Chattoraj and Birdi, 1984) (due to mixed monolayer formation). It has been found that SDS+$Cl_2H_{25}5OH$ and some other additives make *liquid-crystalline* structures at the surface. This leads to a stable foam (and liquid-crystalline structures) (Friberg, 1976). In fact, one deliberately uses SDS in technical formulations, with some (less than 1%) $C_{12}H_{25}OH$, to enhance the foaming properties. The foam drainage, surface viscosity, and bubble-size distributions have been reported for different systems consisting of detergents and proteins. Foam drainage was investigated by using an incident light interference microscope technique.

Furthermore, in the fermentation industry, where foaming is undesirable, the foam is generally caused by proteins. Since mechanical defoaming is expensive, due to the high energy required, *antifoam agents* are generally used, though they are not desirable in some systems, such as food products. In addition, the antifoam agents deteriorate the gas dispersion due to increased coalescence of the bubbles. It has been known for a long time that foams are stabilized by proteins, and that these are dependent on pH and electrolyte. The high foaming capacity is explained by the stability of the gas–liquid interface due to the denaturation of protein, especially due to their strong adsorption at the interface, which gives rise to the stable monomolecular films at the interface. Foam stability is caused by film cohesion and elasticity. The effect of electrolytes and alcohol was investigated. A good correlation was found between the adsorption kinetics and the foaming properties.

Foam structure: Foam as TLF has a very fascinating structure. If two bubbles of the same radius come into contact with each other, this leads to the formation of a contact area and subsequently to the formation of one large bubble. This leads to the following considerations (Wilson, 1989; Birdi, 2016):

1. Two bubbles of same radius
2. Two bubbles touch each other and form a contact area ($=dA_c$)
3. Formation of only one bubble

In stage 2, the energy of the system is higher than in 1, since the system has formed a contact area (A_c). The energy difference between 2 and 1 is $\gamma\,A_c$. When stage 3 is reached, there will be a decrease in total area by 41% (i.e., the sum of the areas of two bubbles is larger than that of one bubble). This means that system 3 is at a lower energy state than the initial state 1. When three bubbles come into contact, the equilibrium angle will be 120°. The angle of contact relates to a system's equilibrium state. If four bubbles are attached to each other, then the angle will, at equilibrium, be 109°28′.

7.3.2 FOAM FORMATION AND SURFACE VISCOSITY

The surface and bulk viscosities not only reduce the draining rate of the lamella, but they also help in restoration against mechanical, thermal, or chemical shocks. The highest foam stability is associated with appreciable η_s and yield value. For example, the over-foaming characteristics of beer (*gushing*) has been the subject of many investigations. The extreme case of gushing is when a beer, on opening, starts to foam out of the bottle, and, in some cases, empties the whole bottle. The relationship between surface viscosity, η_s, and gushing was reported by various investigators. The various factors described for the gushing process were pH, temperature, and metal ions, which could lead to protein denaturation. The stability of a gas (i.e., N_2, CO_2, air) bubble in a solution depends on its dimensions, besides other parameters (Scheludko, 1966). A bubble with a radius greater than a *critical magnitude* will continue to expand indefinitely and degassing of the solution will take place. Bubbles with a radius equal to the critical value will be in equilibrium, while bubbles with a radius less than the critical value would be able to redissolve in the bulk liquid. The magnitude of the *critical radius*, R_{cr}, varies with the degree of saturation of the liquid, that is, the higher the level of supersaturation, the smaller the R_{cr}. The work, W_{bubble}, required for the formation of a bubble of radius R_{cr} is given by (La Mer,1962; Birdi, 2016)

$$W_{bubble} = \left(16\pi\gamma^3\right)/\left(p_{in} - p_{bulk}\right) \tag{7.7}$$

where:
  γ is the surface tension of beer
  p_{in} is the pressure inside the bubble
  p_{bulk} is the pressure in the bulk liquid

It has been suggested that there is nothing unusual in the stability of the beer and, although carbon dioxide is far from an ideal gas, empirical work supports this conclusion. A possible connection between η_s and gushing has been reported. Nickel ion, a potent inducer of gushing, has been reported to give rise to a large increase in the η_s of beer. Other additives besides Ni, such as Fe or humulinic acid, which cause gushing, have also been reported to give a large increase in η_s. On the other hand, additives such as ethylenediamine acetic acid (EDTA: a chelating agent), which are reported to inhibit gushing, have been reported to decrease the η_s of beer. This relation between η_s and gushing suggests that an efficient gushing inhibitor should be very surface active in order to be able to compete with gushing promoters, but is incapable of forming rigid surface layers (i.e., high η_s). An unsaturated fatty acid, such as linoleic acid, is a potent gushing inhibitor, since it destabilizes the surface films. The surface viscosity, η_s (g s^{-1}), was investigated by the oscillating-disc method. It was found that low η_s (0.03–0.08 g s^{-1}) beer surfaces presented nongushing behavior. Beers with high η_s (2.3–9.0 g s^{-1}) presented gushing behavior.

7.3.3 ANTIFOAMING AGENTS

Foam is a typical example of a system that can be both desirable and undesirable, depending on the phenomenon under analysis (Scheludko, 1966; Adamson and Gast, 1997; Birdi, 2016). One finds foaming to be undesirable in such cases as dishwashing, or in many industries, such as wastewater treatment. The main criteria for antifoaming molecules is that they exhibit following characteristics:

- Do not form mixed monolayers
- Reduce surface viscosity, η_s (thus destabilizing the foam films)
- Have a low boiling point (such as ethanol)

7.4 WASTEWATER PURIFICATION (BUBBLE FOAM METHOD)

The biggest challenge mankind is facing is the need to supply pure drinking water worldwide. The world population increase (by a factor of 4 from 1900 to 2000) is much faster than the increase in the availability of pure drinking water supplies. The increased need for water in industrial production also adds a further burden on clean water supplies. The purification of water for households has been developed during the past decades. Pollutants found in wastewater are of different origin and concentration. Solid particles are mostly removed by filtration (or flotation), but the colloidal particles are easily removed by this method. Solute compounds are rather difficult to remove, especially toxic substances with very low concentration. Flotation has been used to great advantage in some cases where sedimentation cannot remove all the suspended particles. The following are some examples where flotation is being used with much success:

- Paper fiber removal in the pulp and paper industry
- Oils, greases, and other fats in food, oil refinery, and laundry wastes
- Clarification of chemically treated waters in potable water production
- Sewage sludge treatment

Many of the industrial wastewaters amenable to clarification by flotation are colloidal in nature, for example, oil emulsions (oil industry, shale oil/gas industry), pulp and paper wastes, and food processing. For the best results, such wastes must be coagulated prior to flotation. In fact, flotation is always the last step in the treatment. In order to aid the effectivity of flotation, one uses surfactants. This leads to lower surface tension and foaming. The latter helps to retain the particles in the foam under flotation.

7.4.1   FROTH FLOTATION (AN APPLICATION OF FOAM) AND BUBBLE FOAM PURIFICATION METHODS

For example, in the case of *hydraulic fracking* technology, one has wastewater, which needs to be treated (Appendix II). In general, wastewater can be treated by various procedures, which will depend on the specific situation. In most cases, multiple treatments are also employed. One of the best-known methods is the *froth flotation* technique for wastewater (Klimpel, 1995; Birdi, 2016; Weber and Clavin, 2012). This procedure is based on using (air or some other gas) bubbles, which adhere to solid particles as found in wastewater. This leads to the froth flotation of solid particles to the surface of water. It was found, many decades ago, that foam or bubbles could be used to purify wastewater in this way. A simple device as used for laboratory froth flotation studies is shown in Figure 7.3.

The bubbles are formed in the sintered glass, as air or other suitable gas (N_2, CO_2, etc.) is bubbled through the solution containing the solid suspension. A suitable flotation agent (a surface-active agent) is added, and the air is bubbled. Surface-active pollutants in wastewater have been removed by bubble film separation methods. In particular, very minute concentrations are easily removed by this method, which is more economical than more complicated methods (such as active charcoal, filtration, and various other chemical methods). However, one must also remove pollutants that may have attached to the SAS (such as dyes, heavy metals, etc.). This method is now commercially available for such small systems as fish tanks, and so on. The principle in this procedure is to create bubbles in a wastewater tank and to collect the bubble foam at the top (Figure 7.4).

Bubbles are blown into the inverted funnel. Inside the funnel, the bubble film is transported away and collected. Since the bubble film consists mostly of

- Surface-active substance
- Water

it is thus seen that even very minute amounts (less than a milligram per liter) of SAS will accumulate at the bubble surface. As shown previously, it would require a large number of bubbles to remove a gram of substance. However, since one can blow thousands of bubbles in a very short time, the method is found to be very feasible.

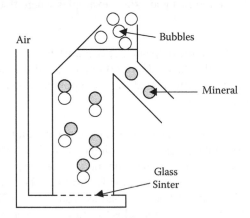

FIGURE 7.3   Apparatus used for froth flotation experiments.

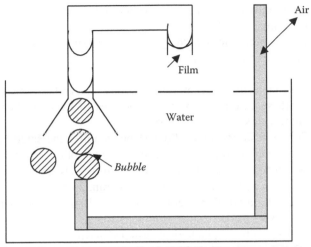

Bubble film formation

FIGURE 7.4   Bubble foam separation method for wastewater purification.

7.5   APPLICATIONS OF SCANNING PROBE MICROSCOPES (STM, AFM, FFM) TO SURFACE AND COLLOID CHEMISTRY

A few decades ago, a big surge in the development of very important techniques became available for surface science and technology: self-assembly structures (micelles, monolayers, vesicles); biomolecules; biosensors; surface and colloid chemistry; nanotechnology. In fact, the current literature shows that these developments continue to grow and supply new levels of information with ever-more sensitive measurement tools.

Typical of all humans, seeing is believing, so the microscope has attracted much interest for many decades. These inventions are based on the principles laid out by the telescope (as invented by Galileo) and the light-optical microscope (as invented by Hooke). Over the years, the magnification and the resolution of microscopes have improved. For man to understand nature, the main aim has been to be able to see atoms or molecules. This goal has now been achieved, and what follows here will explain the latest developments. The ultimate aim of scientists has always been to be able to see molecules while they are active. In order to achieve this goal, the microscope should be able to operate under ambient conditions. All kinds of molecular interactions between a solid and its environment (gas or liquid or solid), initially, can take place only via the surface molecules of the interface. One finds that in geology (rock structures and surfaces, surface forces), there are phenomena that can be investigated by microscopes. It is obvious that when a solid or liquid interacts with another phase, knowledge of the molecular structures at these interfaces is of interest. The term *surface* is generally used in the context of gas–liquid or gas–solid phase boundaries, while the term *interface* is used for liquid–liquid or liquid–solid phases. Furthermore, many fundamental properties of surfaces are characterized by morphology scales of the order of 1–20 nm (1 nm $= 10^{-9}$ m $= 10$ Å (Angstrom $= 10^{-8}$ cm).

It is found from experiments that the general basic issues that should be addressed for these different interfaces are as follows.

- What do the molecules of a solid surface look like, and how are the characteristics of these different to the bulk molecules? In the case of crystals, one asks about the kinks and dislocations
- Adsorption or desorption on solid surfaces requires the same information about the structure of the adsorbates and the adsorption site and configurations. The adsorption site has been found to be selective on some solid surfaces (Adamson and Gast, 1997; Somasundaran, 2015; Birdi, 2003, 2016)
- Solid–adsorbate interaction energy is also required, as is known from the Hamaker theory
- Molecular recognition in biological systems (active sites on the surfaces of macromolecule, antibody–antigen) and biological sensors (enzyme activity, biosensors)
- Self-assembly structures at interfaces
- Semiconductors

Depending on the sensitivity and experimental conditions, the methods of molecular microscopy are many and varied. The applications of these microscopes are also very varied and extensive. For example, they have provided information about crystal structures and the three-dimensional configurations of macromolecules.

The most common application of microscopy is the study of molecules at surfaces. Generally, the study of surfaces is dependent on understanding not only the reactivity of the surface but also the underlying structures that determine that reactivity. Understanding the effects of different morphologies may lead to a process for enhancement of a given morphology, and hence to improved reaction selectivities and product yields. Atoms or molecules at the surface of a solid have fewer neighbors as compared with atoms in the bulk phase, which is analogous to the liquid surface; therefore, surface atoms are characterized by an unsaturated, bond-forming capability and accordingly are quite reactive. Until a decade ago, electron microscopy and some other similarly sensitive methods provided some information about the interfaces, but there were always some limitations inherent in all these techniques. Now, however, due to relentless technical advances, electron crystallography is capable of producing images at resolutions close to those attained by x-ray crystallography or multidimensional nuclear magnetic resonance (NMR). In order to improve on some of the limitations of the electron microscope, newer methods and procedures were needed. The recent scanning probe microscopes (SPM) not only provide a new kind of information, as known from x-ray diffraction, for example, but they also open up new areas of research, in nanoscience and nanotechnology, for example.

The basic operating principle of SPMs (Binnig and Rohrer, 1983; Birdi, 2003, 2016) is essentially to be able to move a tip (at a nm distance) over the substrate surface with a sensor (probe) with molecular sensitivity (nm) in both the longitudinal and height direction (Figure 7.5). This may be compared with the act of sensing with a finger

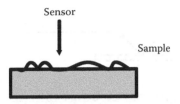

Sensor

Sample

FIGURE 7.5   A typical AFM SPM.

TABLE 7.1
Areas of Application for the STM and AFM

Lipid monolayers (as Langmuir–Blodgett films)
Different layered substances on solids
Self-assembly structures at interfaces
Solid surfaces
Langmuir–Blodgett films
Thin film technology
Interactions at surfaces of ion beams/laser damage
Nano-etching and lithography, nanotechnology
Semiconductors
Mineral (shale) surface morphology
Metal surfaces (roughness)
Micro-fabrication techniques
Optical and compact discs
Ceramic surface structures
Catalyses
Surface adsorption (metals, minerals)
Surface manipulation by STM/AFM
Polymers
Biopolymers (peptides, proteins, DNA, cells, viruses)
Vaccines

---

over a surface or more akin to the old-fashioned record player with a metallic needle (a probe for converting mechanical vibrations to music sound) moving on a vinyl record.

Scanning probe microscopy was invented a few decades ago (by Binnig and Rohrer, who won the Nobel Prize in 1986). The scanning tunneling microscope (STM) was based on scanning a probe (metallic tip) since it is a sharp tip moving just above the substrate while monitoring some interaction between the probe and the surface. The tip is controlled to within 0.1 Å.

In an SPM, the various interactions between the tip and the substrate are as follows (Table 7.1):

- STM: The tunneling current between a metallic tip (ca. 0.2 mm) and a conducting substrate, which are in very close proximity, but not actually touching. This is controlled by piezo motors in a stepwise method.

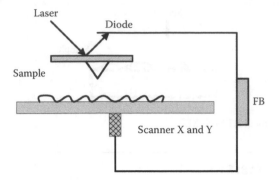

FIGURE 7.6 A schematic drawing of the sensor (tip/cantilever/optical/magnetic device) movement over a substrate in the x/y/z direction with nanometer sensitivity (controlled by piezo motor) (at solid–gas or solid–liquid interface).

- AFM: The tip is brought closer to the substrate while the Van der Waals force is monitored. At a given force, the piezo motor controls this setting while the surface is scanned in x–y direction.
- FFM: This is a modification of AFM, where force is measured (Birdi, 2003) (Figure 7.6).

The most significant difference between SPM and x-ray diffraction studies has been that the former can be carried out both in the air and in water (or in any other fluid). Corrosion and similar systems have been investigated using STM. A plastic material covers the tip, and this allows one to operate the STM in a fluid environment. STM has been used to study the molecules adsorbed on solid surfaces. Adsorbed lipid molecules (Langmuir-Blodgett [LB] films have been extensively investigated by both STM and AFM). The effect of chain length and other structures has been investigated. AFM has been used to study surface molecules under different conditions.

7.5.1 MEASUREMENT OF ATTRACTIVE AND REPULSIVE FORCES (BY AFM)

7.5.1.1 Shale Rock and Other Solid Surfaces

As mentioned in Chapter 6, as two bodies approach with a distance of separation of molecular dimensions, both short-range (vdW) and long-range (Coulombic forces) interaction forces exist (Bockris et al., 1980; Kortum, 1965; Birdi, 2010b, 2016). AFM has been reported to be a very useful technique to study these interactions (Birdi, 2003; Javadpour et al., 2012).

Direct interactive force studies by AFM between surfaces: AFM has allowed scientists to be able to study molecular forces between molecules at very small (almost molecular size) distances (Birdi, 2003; Javadpour et al., 2012). AFM is a very attractive and sensitive tool for such measurements. It can be applied to study samples *in situ*, without any treatment. The scanning microscopes can also operate under fluids (Birdi, 2003). This makes it possible to make measurements under varying compositions of the separating fluid (such as additives [salts, pH, etc.]). The most

Silica
sphere

FIGURE 7.7 AFM sensor with SiO_2 sphere.

important study was where the colloidal forces as a function of pH of SiO_2 immersed in aqueous phase was reported using AFM (Hu and Bard, 1998; Birdi, 2003). The force between an SiO_2 sphere (ca. 5 mm diameter) and a chromium oxide surface in an aqueous phase of sodium phosphate were measured (pH from 3 to 11). The SiO_2 sphere was attached to the AFM sensor as shown in Figure 7.7. In the literature, one finds that different solids (other than SiO_2) have also been investigated by AFM.

The magnitude of the deflection of the cantilever has been correlated to the force acting between the tip and the surface (Hook's law):

$$F_{afm} = k_{cant} D_{cant} \qquad (7.8)$$

where:
F_{afm} is the force acting between the cantilever (or SiO_2 sphere) and the surface
k_{cant} is the constant of stiffness of the cantilever
D_{cant} is the deflection

These data showed that the isoelectric point (IEP) of SiO_2 was around pH 2, as expected. The binding of phosphate ions to a chrome surface was also estimated as a function of pH and ionic strength (Birdi, 2003b). The force curves were related to the degree of adsorption of octa-decyltrichloro-silane and Si probe (Hu and Bard, 1998). Furthermore, both STM and AFM have been used to investigate corrosion mechanisms of metals exposed to aqueous phase. Since both STM and AFM can operate under water, this gives rise to a variety of possibilities. Extensive analyses have been reported in the literature (Birdi, 2003).

AFM studies of shale rocks have been recently reported (Javadpour et al., 2012). In these studies, the deflection between the cantilever and shale surface was investigated (in an air medium). These studies could distinguish between organic surface (kerogen) and inorganic areas. It is also reported that adsorption of organic molecules (such as alkanes) can be studied by SPM (McGonigal et al., 1990; Castro et al., 1998; Birdi, 2003).

Molecular domains have been observed using the AFM analyses of LB films of collapsed state (Birdi, 1997, 2003). For example, cholesterol films showed *half-butterfly* shaped domains (each domain consisting of 10^7 molecules). This quantity was estimated from the following data. The height of the domains was 90 Å, which corresponds with six layers of the cholesterol molecule (length of molecule is found to be 15 Å from molecular models). The AFM image analysis is capable of calculating the area of the image. Since one knows the magnitude of the area/molecule of the cholesterol molecule (40 Å^2), this analysis can be carried out. In comparison to macro-domains, the latter domains thus are measured as three-dimensional by AFM. This indicates that such nanostructures can be investigated by using AFM. As compared with ordinary microscopes, the SPM provides 2-D and 3-D images. The 3-D images allow one to see molecules with different diameters (mixed polymer systems) (Birdi, 2003).

8 Emulsions and Microemulsions

Oil and Water Mixtures

8.1 INTRODUCTION

In this chapter, special applications of surface chemistry principles pertaining to oil and water phases will be considered. Oil and water do not mix if shaken. As is well known, if one shakes oil and water, oil breaks up into small drops (about a few mm diameters) but these drops join together rather quickly to return to their original state (as depicted here).

- Step I: Oil phase and water phase
- Step II: Mixing
- Step III: Oil drops in water phase
- Step IV: After a short time
- Step V: Oil phase and water phase

One finds oil–water systems in all kinds of technical and biological phenomena (cosmetics, foods, pharmaceutical, building, paints, oil industry, wastewater treatment, biology). However, one finds that oil and water can be dispersed with the help of suitable *emulsifiers* (surfactants) to give *emulsions* (Becher, 2001; Dickenson, 1992; Friberg et al., 2003; Birdi, 2007). This is well known in emulsions, such as mayonnaise, found in the household. The basic reason is that the interfacial tension (IFT) between oil and water is around 50 mN m^{-1}, which is high and leads to the formation of large oil drops. On the other hand, with the addition of suitable emulsifiers, one can reduce the value of the IFT to very low values (much less even than 1 mN m^{-1}). Emulsion formation means that oil drops remain dispersed for a given length of time (up to many years). The stability and the characteristics of these emulsions are related to the area of application.

Emulsions are mixtures of two or more immiscible substances (Birdi, 2014). Some everyday common examples are milk, butter (fats, water, salts), margarine, mayonnaise, and similar. In butter and margarine, the continuous phase consists of lipids. These lipids surround the water droplets (water-in-oil emulsion). All technical emulsions are prepared by some convenient kind of mechanical agitation or mixing.

8.1.1 EMULSIONS AND HYDRAULIC FRACKING

The technology of fracking methods has developed rapidly during the last decade. In some applications, different emulsion-based fracking fluids have been reported. Most of these fluids were oil–water emulsions. The aim has been to reduce the amount of water used in the process. In one case, carbon monoxide (CO) dispersed in aqueous alcohol gel was used. CO foams have also been used (Gupta and Hlidek, 2009; Shabro, 2013). These emulsions are being investigated in an attempt to reduce the amount of water used in such processes. Certain reservoir formations have the potential to retain even the smallest amounts of water contained in foams. In these formations, a 40% methanol aqueous system yielded very good production results in several Canadian gas formations (Appendix II). These studies indicated the following advantages, as regards reduction in water:

- Amount of water reduced or completely eliminated
- Reduction in amount of additives used
- Increase in production

8.2 STRUCTURE OF EMULSIONS

Emulsions are some of the most important application areas of surface-active compounds. These systems are generally categorized into three different kinds:

- Emulsions
- Microemulsions
- Liquid crystals (LC) and lyotropic LCs

Emulsions are systems where one needs to apply both water and oil to an application. This may be a skin treatment (cosmetics or foods [milk, butter]), or shoe polish, or similar. In other words, one can apply both components (water and oil), which do not mix simultaneously. This also allows one to perform functions that are dependent on water or oil. In most emulsion systems, two liquid phases are involved. Though, in some complex systems, such as milk or butter, one may have more than two main components.

This is explained by gathering information about the IFT, as well as the solubility characteristics of surface-active substances (SAS) needed to stabilize emulsions.

Microemulsions are microstructured mixtures of oil–water–emulsifiers–other substances. Microemulsions are found to differ in many ways from the ordinary emulsion structure. LCs are substances that exhibit special melting characteristics. Furthermore, some mixtures of surfactant–water–cosurfactant may also exhibit lyotropic LC properties. The emulsion technology is basically thus concerned in preparing mixtures of two immiscible substances:

- Oil
- Water

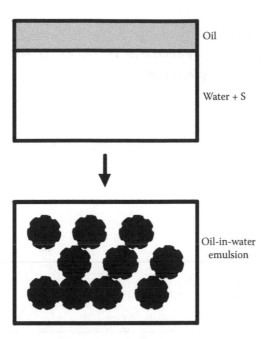

FIGURE 8.1 Mixing of oil–water (a) or oil–water + surfactant (b) by shaking.

by adding suitable surface-active agents (emulsifiers, cosurfactants, polymers). When a surface-active substance is added to a system of oil–water, the magnitude of IFT decreases from 50 mN m^{-1} to 30 (or lower [less than 1]) mN m^{-1}. This leads to the observation that on shaking an oil–water system, the decreased IFT leads to smaller drops of the dispersed phase (oil or water). The smaller drops also lead to a more stable emulsion. Depending on the surfactant used, one will obtain oil in water (O/W) or water-in-oil (W/O) emulsion. These experiments, where oil–water or oil–water + surfactant are shaken together, are shown in Figure 8.1.

These emulsions are all opaque, since they reflect light. Some typical oil–water IFT values are given in Table 8.1.

These data show certain trends. The decrease in IFT is much smaller with the decrease in alkyl chain in the case of alkanes than in alcohols.

8.2.1 Oil–Water Emulsions

Emulsions are among the most important structures that are prepared specifically for a given application. For example, in skincare, day cream has different characteristics and ingredients than night cream. One of the main differences in emulsions is whether oil droplets are dispersed in the water phase or water drops are dispersed in the oil phase. One can determine this by measuring the conductivity, since it is higher for O/W than for W/O emulsion. Another useful property is that O/W will dissolve water while W/O will not. This thus shows that one will choose W/O or O/W depending on the application area. In the case of skin emulsions, the type is very importance.

TABLE 8.1

Magnitudes of Interfacial Tensions of Different Organic Liquids against Water (20°C)

Oil Phase	IFT (mN/m)
Hexadecane	52
Tetradecane	52
Dodecane	51
Decane	51
Octane	51
Hexane	51
Benzene	35
Toluene	36
CCl_4	45
CCl_3	32
Oleic acid	16
Octanol	9
Hexanol	7
Butanol	2

- Oil-in-water emulsions: The main criteria for an O/W emulsion will be that if one adds water to it, it will be miscible. Also, after water evaporates, the oil phase will be left behind. Thus, if one needs an oil phase on the substrate (such as skin, metal, wood), then one should use an O/W type emulsion.
- Water-in-oil emulsions: The criterion for an W/O emulsion is that it is miscible with oil. That means that if one adds the emulsion to some oil, then one obtains a new but diluted W/O emulsion. In some skin creams, W/O-type emulsions are preferred (especially if an oil-like feeling after application is needed).

8.2.2 HLB Values of Emulsifiers

The emulsifiers used exhibit varying solubility in water (or oil). This will thus have consequences on the emulsion. Let us consider a system where we have oil and water. If we add an emulsifier to this system, then it will be distributed both in the oil and the water phase. The degree of solubility in each phase will depend on its structure and hydrophilic–lipophilic balance (HLB) character. The emulsifiers used in making emulsions are characterized with regard to the molecular structure. The amphiphile molecules consist of HLB characteristics. Thus, each emulsifier that may be needed for a given system (for example if one needs an O/W or W/O emulsion) will need a specific HLB value. The data in Table 8.2 give a rough estimation of the HLB needed for a given system of emulsion. In general, it is expected that if the emulsifier dissolves in water, then, on adding oil, an oil-in-water emulsion is obtained. Conversely, if the emulsifier is soluble in the oil, then, on adding water, a water-in-oil emulsion is obtained.

TABLE 8.2
HLB Values and Emulsion Type

Emulsifier Solubility	HLB	Application in Water
Low solubility	4–8	W/O
Soluble	10–12	Wetting agent
High solubility	14–18	O/W

W/O emulsions are formed by using HLB values between 3.6 and 6. This suggests that one generally uses emulsifiers, which are soluble in the oil phase. O/W emulsions need HLB values of about 8–18. This HLB criterion is only a very general observation. However, it must be noticed that HLB values alone do not determine the emulsion type (or stability). Other parameters, such as temperature, the properties of the oil phase, and electrolytes in the aqueous phase also affect the emulsion. The HLB values have no relation to the degree of emulsion stability. The HLB values of some surface-active agents are given in Table 8.3.

The HLB values decrease as the solubility of the surface-active agent decreases in water. The solubility of cetyl alcohol in water (at 25°C) is less than a milligram per liter. It is thus obvious that in any emulsion, cetyl alcohol will be present mainly in the oil phase, while SDS will be mainly found in the water phase. The empirical HLB values are found to have significant use in applications in emulsion technology.

TABLE 8.3
HLB Values of Some Typical Emulsifiers

SAA	HLB
Na-lauryl sulphate	40
Na-oleate	18
Tween80(sorbitan monooleate EO20)	15
Tween81(sorbitan monooleate EO6)	10
Ca-dodecylbenzene sulfonate	9
Sorbitan monolaurate	9
Soya lecithin	8
Sorbitan monopalmiate	7
Glycerol monolaurate	5
Sorbitan monostearate	5
Span80(sorbitan monooleate)	4
Glycerol monostearate	4
Glycerol monooleate	3
Sucrose distearate	3
Cetyl alcohol	1
Oleic acid	1

It was shown that the HLB is related, in general, to the distribution coefficient, K_D, of the emulsifier in the oil and water phases (Birdi, 2016):

$$K_D = C(\text{water})/C(\text{oil}) \tag{8.1}$$

where:
C(water) is the equilibrium molar concentrations of the emulsifier in the water phase
C(oil)   is the equilibrium molar concentrations of the emulsifier in the oil phase

The thermodynamics of this equilibrium are used to correlate HLB to K_D, as follows (Birdi, 2016):

$$(\text{HLB}-7) = 0.36 \ln(K_D) \tag{8.2}$$

Based on these thermodynamic relations, one could then suggest the relation between HLB and emulsion stability and structure.

The HLB values can also be estimated from the structural groups of the emulsifier (Table 8.4) (Birdi, 1997, 2016). Table 8.4 can be useful in those cases where one needs to estimate the HLB value.

In the food industry, one finds many applications of food emulsifiers. These emulsifiers must satisfy special requirements (e.g., toxicity) in order to be useful in the food industry. It is thus suggested that in shale fracking, similar toxicity restrictions are acceptable and possible. One determines the toxicity from animal tests. The test determines the amount of a substance that causes 50% (or more) of the test animals

TABLE 8.4
HLB Group Numbers

Group	Group Number
Hydrophilic	
$-SO_4Na$	39
$-COOH$	21
$-COONa$	19
Sulfonate	11
Ester	7
$-OH$	2
Lipophilic	
$-CH$	0.5
$-CH_2$	0.5
$-CH_3$	0.33
$-CH_2CH_2O-$	39

to die (lethal dosage; LD50). It is thus obvious that food emulsions are subject to much stricter controls (Friberg, 2003).

8.2.3 Methods of Emulsion Formation

If one shakes oil and water, the oil breaks up into drops. However, these will quickly coalescence and return to their original state of two different phases. One also observes that the more one shakes, the more drops reduce in size. In other words, the energy put into the system makes the drops smaller in size. Emulsions are made by different procedures. These can include mechanical agitation and other methods. Industry uses state-of-the-art emulsion technology (Sjoblom, 2001; Friberg, 1976; Holmberg, 2002; Birdi, 2003). Therefore, a vast literature about the methods used exists for any specific emulsion. In a simple case, an emulsion may be based on three necessary ingredients: water, oil, and emulsifier. In other words, one needs to determine in which weight proportions one needs to mix these substances in order to obtain an emulsion (at a given temperature) to be stable (or to achieve maximum stability). This may be more conveniently carried out in a phase study in the triangle. The micellar region exists on the water–surfactant line (Figure 8.2).

Near the surfactant region, one finds the *crystalline* or *lamellar* phase. This is the region in which one finds hand soaps. Ordinary hand soap is mainly salts of fatty acid (coconut oil, fatty acids, or mixtures) (85%) plus water (15%) and some salts and so on. X-ray analyses have shown that the crystalline structure consists of a series of a layer of soap separated by a water layer (with salts). The hand soap is produced by extruding under high pressure. This process aligns the lamellar crystalline structure lengthwise. It is further found that complex structures are present in the other regions in the phases (Figure 8.2). The diagram is strongly dependent on temperature.

In practice, what one does is as follows. A suitable number (over 50) of test samples are prepared by mixing each component in varying weights to represent a suitable number of regions (around 50 samples). The test samples are mixed under rotation in a thermostat over a few days to reach equilibrium. The test samples are centrifuged

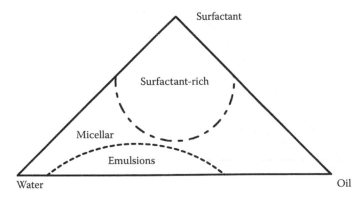

FIGURE 8.2 Different phase equilibria in a water–surfactant (emulsifier)–oil mixture system.

and the phases are analyzed. From these analyses, the phases are determined. The phase structure is investigated by using a suitable analytical method.

It is obvious that studies of multicomponent systems such as these will lead to very large numbers of phases. However, by analyzing some typical systems, one finds that there are some trends that can be used as guidelines. For example, another very well-investigated system consists of (Friberg, 1976; Birdi, 2016)

- Water
- Potassium caprate (K-caprate)
- n-Octanol

The phases were determined as indicated in Figure 8.3.

The system is a very useful example to understand what phase equilibria are involved when three components are mixed. Some characteristics are noticeable in this system, which point out the significance of ratios between KC:O. For example, the aqueous phase region is extrapolated to 1 mol octanol:2 mol K-caprate. This shows that the 1:2 ratio dominates the phase region. It has been found from other studies (such as monolayers on water films of lipids) that such mixtures are indeed found. The three-phase region is extrapolated to show that 1 mol octanol:1 mole KC is the ratio. In a much simplified description here, one thus finds that in such complicated phase equilibria, some simple molecular ratios indicate the phase boundaries. Thus, in general, one may safely conclude that these *molecular ratios* will be useful when working with emulsions. The observation that exact ratios exist between different components at the phase lines suggests that some kind of molecular aggregates are formed. These correspond to the formation of some liquid–crystalline structures. Much confirmation on these molecular aggregates has been found from monolayer studies of mixed films spread on water (Birdi, 1984, 1989; Soltis et al., 2004). A similar conclusion was reached when investigating microemulsions (Chapter 8).

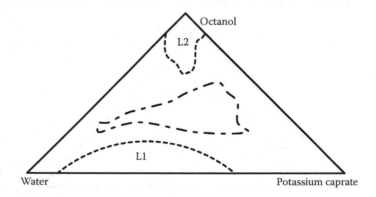

FIGURE 8.3 Phase diagram for the system K-caprate (PK) + water + n-octanol (22°C). All compositions are given in weight % (L1 = micellar phase; L2 = reverse micelle; H1 = hexagonal LC phase).

O:W

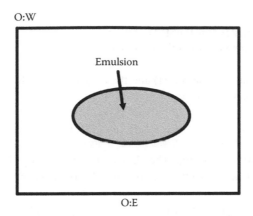

FIGURE 8.4 Emulsion region based on the ratio of oil (O):water (W) vs. oil (O): emulsifier (E).

Furthermore, in practice, one needs to prepare a given emulsion with some specified range of ratio between oil and water. In these cases, one may find it more useful to study mixtures of oil (O)–water (W)–emulsifier (E), as plots of ratios (Figure 8.4). The region of most suitable emulsion can be determined by studying varying mixtures.

8.3 EMULSION STABILITY AND ANALYSES

The stability of emulsions is dependent on various parameters (size of drops, interactions between drops). These different parameters are described in the following.

Emulsion drop size analyses: Since the stability and other characteristics (such as viscosity and appearance) are known to be related to the drop size, one needs to measure these. The following commercial instruments are useful for such analyses:

- Coulter counter: This is the most common type where one simply counts the number of particles or drops passing through a well-defined hole. A signal is produced, which corresponds to the size of the particle.
- Light-scattering: Laser light-scattering instruments are very advanced for particle size distribution analyses. The laser light is scattered by the small dispersed particles or drops. The latter is known to be dependent on the radius of the particle.
- *Emulsion stability*: Emulsions are stable as long as the drops are separated from each other. Flocculation of an emulsion or dispersion takes place upon collision of the droplets, which is related to Brownian motion, convective stirring, or gravitational forces. Any emulsion can be separated into an oil-and-water phase by suitable centrifugation treatment.

The dispersion force of attraction between two different bodies (i and j) (molecules, particles, drops), E_{ij}, is dependent on the following parameters:

$$E_{ij} = H_{ij} / (12 \Pi R^2) \qquad (8.3)$$

where:

   H_{ij} is the Hamaker constant for i and j

   R   is the distance between the particles

Since in emulsions, one has oil (1) and the continuous medium water (2), then the expression for E_{121} is found to be

$$E_{121} = \left(aH_{121}\right)/\left(12R\right) \tag{8.4}$$

where a is the size of the oil drop. The Hamaker constant, H_{121}, is found to be related to the dispersion surface tension, γ_{LD}, such that for oil/water emulsion:

$$H_{121} = 3\times10^{-14}/\varepsilon2 \left(\gamma_{LD}0.5 - \gamma_{2D}0.5\right)^{11/6} \tag{8.5}$$

where:

   γ_{LD} (30 mN/m) is the dispersion surface tension of oil

   γ_{2D} (22 mN/m) is the dispersion surface tension of water

From these equations, we find that if oil/water

$$\gamma_{LD} = 30 \text{ mN/m}; \varepsilon_2 = 1.77 \tag{8.6}$$

then H_{121} is equal to 1.1×10^{-14} ergs. For drops of size equal to 1 μm ($a = 5 \times 10^{-5}$ cm), then E_{121} is equal to almost $k_B T$ (4×10^{-14} ergs, at 298 K (25°C)). The magnitude of H_{121} has been shown to be always positive, which suggests that in two-phase systems (such as oil–water) the particles will always be attracted to each other. This means that even air bubbles will attract each other, as also found from experiments. A linear relation is found between $H_{121}^{6/11}$ and γ_{LD}, as expected from Equation 8.5. Experimental values of A_{121}, as determined from flocculation kinetics, showed that this agreed with the theoretical relation.

8.3.1 Electrical (Charge) Emulsion Stability

There are systems where the emulsifier carries a charge that imparts specific characteristics to the emulsion. A double layer will exist around the oil droplets in an O/W emulsion. If the emulsifier is negatively charged, then it will attract positive counter-ions while repelling negative charged ions in the water phase. The change in potential at the surface of oil droplets will be dependent on the concentration of ions in the surrounding water phase.

The state of stability under these conditions can be qualitatively described as follows: As two oil droplets approach each other, the negative charge gives rise to repulsion. The repulsion will take place within the electrical double-layer region. It can thus be seen that the magnitude of the double-layer (EDL) distance will decrease if the concentration of ions in the water phase increases. This is because the electrical double-layer (EDL) region decreases. However, in all such cases where two bodies come closer, two different kinds of forces exist, which must be considered:

$$\text{Total force} = \text{repulsion forces} + \text{attraction forces}$$

The nature of the total force thus determines whether

- The two bodies will stay apart
- The two bodies will merge and form a conglomerate

This is a very simplified picture, but a more detailed analysis has been presented in the current literature. The attraction force arises from Van der Waals forces. The kinetic movement will finally determine whether the total force can maintain contact between the two particles.

Different processes are involved in the stability and characteristics. The various processes are as follows:

8.3.2   CREAMING OR FLOCCULATION OF DROPS

This process is described in those cases where oil drops (in the case of oil–water) cling to each other and grow in large clusters. The drops do not merge into each other. The density of most oils is lower than that of water. This leads to the fact that instability in the oil-drop clusters rises to the surface (Ivanov and Kralchevsky, 1997; Birdi, 2016). One can reduce this process by

1. Increasing the viscosity of the water phase and thereby decreasing the rate of movement of the oil drops
2. Decreasing the IFT and thus the size of the oil drops

The ionized surfactants will stabilize O/W emulsions by imparting surface EDL.

The degree of stability of any emulsion is related to the rate of coagulation of two drops (O/W: oil drops; W/O: water drops) to form one large drop. This process means that two oil drops in an O/W emulsion come close together and if the repulsion forces are smaller than the attraction forces, only then will the two particles meet and fuse into one larger drop. In the case of charged drops, an EDL will be present around these drops (Adamson and Gast, 1997; Birdi, 2010a). A negatively charged oil drop (charge arising from the negatively charged emulsifier) will strongly attract positively charged counter-ions in the surrounding bulk aqueous phase. At a close distance from the surface of a drop, the distribution of charges will be very much changing. While at a very large distance, there will be electrical neutrality, as there will be an even number of positive and negative charges. Electrostatic repulsion exists between the two negatively charged drops, which would exhibit strong repulsion even at large distances (many times the size of the particle). The shape of the EDL curve will be dependent on the negative and positive charge distribution. It is easily seen that if the concentration of counter-ions increases, then the magnitude of EDL will decrease and this will decrease the maximum of the total potential curve. The stability of emulsions can thus be increased by decreasing the counter-ion concentration. Another important emulsion stabilization technique is achieved by using polymers. The large polymer molecules adsorbed on solid particles will exhibit repulsion at the surface of the particles. The charged polymers will thus also give additional charge–charge repulsion.

8.4 ORIENTATION OF AMPHIPHILE MOLECULES AT OIL–WATER INTERFACES

Currently, there is no method available in the literature by which one can directly determine the orientation of molecules of liquids at interfaces. Molecules are situated at interfaces (e.g., air–liquid, liquid–liquid, solid–liquid) under asymmetric forces. Some studies have been carried out to obtain information about molecular orientation from surface tension studies of fluids (Birdi, 1997). It has been concluded that interfacial water molecules in the presence of charged amphiphiles are in a tetrahedral arrangement similar to the structure of ice. Other studies of alkanes near their freezing point had indicated that surface tension changes in abrupt steps. X-ray scattering of liquid surfaces indicated similar behavior (Wu et al., 1993). However, it was found that lower-chain alkanes (hexadecane: $C_{16}H_{34}$) did not show this behavior. The crystallization of $C_{16}H_{34}$ at 18°C shows an abrupt change due to the contact angle change at the liquid—Pt plate interface (Birdi, 1997). It was found that in comparison with $C_{16}H_{34}$–air interface one observes super cooling (to ca. 16.4°C). Each data point corresponded to 1 s, thus the data showed that crystallization is very abrupt. High-speed data ($\ll 1$ s) acquisition is needed to determine the kinetics of transition. This kinetic data would add more information about molecular dynamics at interfaces.

8.5 MICROEMULSIONS (OIL–WATER SYSTEMS)

As mentioned in Section 8.2, ordinary emulsions as prepared by mixing oil–water–emulsifier are thermodynamically unstable. In other words, such an emulsion may be stable over a long time, but ultimately, it will separate into two phases (oil phase and aqueous phase). All such emulsions can be separated into two phases, that is, oil phase and water phase, by centrifugation. These emulsions are opaque, which means that the dispersed phase (oil or water) is present in the form of large droplets (over μm and thus visible to the naked eye).

A microemulsion is defined as a thermodynamically stable and a clear isotropic mixture of water–oil–surfactant–cosurfactant (in most systems, it is short-chain alcohols). The cosurfactant is the fourth component, which gives rise to the formation of very small aggregates or drops that make the microemulsion almost clear.

Microemulsions are also, therefore, characterized as microstructured, thermodynamically stable mixtures of water:oil:surfactant:additional components (such as cosurfactants, etc.).

The study of microemulsions has shown that they are one of the following types.

- Microdroplets of oil-in-water or water-in-oil
- Bicontinuous structure

Emulsifiers will be found in both these phases. On the other hand, in systems with four components (Figure 8.5) consisting of oil–water–detergent–cosurfactant, there

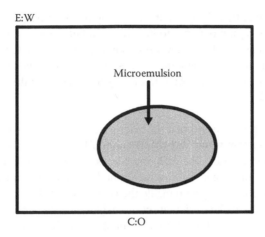

FIGURE 8.5 Four-component system: oil (O)–water (W)–emulsifier (E)–cosurfactant (S) (ratio of O:S vs. S:W).

exists a region where a clear phase (i.e., microemulsion) is found. Microemulsions are thermodynamically stable mixtures. The IFT is almost zero. The size of the drops is very small, which makes microemulsions seem clear. It has also been suggested that microemulsions may consist of bicontinuous structures. This sounds more plausible in these four-component microemulsion systems. It has been suggested that microemulsions may be compared with swollen micelles (that is, if one solubilizes oil in micelles). In such isotropic mixtures, short-range order between the droplets exists. Since it has been found from extensive experiments that not all mixtures of water–oil–surfactant–cosurfactant give rise to a microemulsion, some studies have tried to predict the molecular relationship.

Microemulsions have been formed by one of following procedures:

- Oil–water mixture is added to a surfactant. To this emulsion, one keeps adding a short-chain alcohol (with 4–6 carbon atoms) until a clear mixture (microemulsion) is obtained. It is thus obvious that microemulsion will exhibit very special properties, quite different from those exhibited by the ordinary emulsions. The microdrops may be considered as large micelles.

A very typical microemulsion, extensively investigated, consists of a mixture of SDS + C_6H_6 + water + cosurfactant (C_5OH or C_6OH).

The phase region is determined by mixing various mixtures (approximately 20 samples) and allowing the system to reach equilibrium under controlled temperature. From the literature, one finds the following recipe (Birdi, 1982; 2016):

- Mix 0.0032 mol (0.92 g) SDS (mol. wt. of SDS ($Cl_2H_{25}SO_4Na$) = 288) with 0.08 mol (1.44 g) water and add 40 ml of C_6H_6. This mixture is mixed by vigorous stirring and one gets a creamy emulsion. Under stirring, to this three-component mixture, a cosurfactant (such as: $C_5H_{11}OH$ or $C_6H_{13}OH$) is added slowly until a clear system consisting of a microemulsion is obtained. The stability region is found to be a relation between surfactant–water and surfactant–alcohol. This shows that some kind of structure (at molecular level) is responsible—that a liquid crystal structure is indeed involved. The size of oil droplets is under a micrometer and therefore the mixture is clear (Birdi, 1982).

These data clearly indicated that the microemulsion phase was formed at certain fixed ratios of surfactant:water and cosurfactant:oil.

It is important to consider the different stages when one proceeds to microemulsions from macroemulsions. It was mentioned earlier that surfactant molecules orient with the hydrophobic group inside the oil phase, while the polar group orients toward the water phase. The orientation of surfactants at such interfaces cannot be measured by any direct method. Although much useful information can be obtained from monolayer studies of air–water or oil–water interfaces.

At present, it is generally accepted that the theoretical basis of a given system is not easy to predict with the microemulsion recipe. However, some suggestions have been put forward, which one may summarize as follows:

- The HLB value of the surface needs to be determined (for deciding the O/W or W/O type)
- The phase diagram of the water–oil–surfactant (and cosurfactant) needs to be determined
- The effect of temperature is found to be very crucial
- The effect of added electrolytes is of additional importance

The phase equilibria of a microemulsion were reported. The phase behavior of a microemulsion formed with food-grade surfactant sodium bis-(2-ethylhexyl) sulfosuccinate (AOT) was studied. Critical microemulsion concentration ($c\mu c$) was deduced from the dependence of pressure of cloud points on the concentration of surfactant AOT at constant temperature and water concentration. The results show that there are transition points on the cloud point curve in a very narrow range of concentration of surfactant AOT. The transition points were changed with the temperature and water concentration. These phenomena show that lower temperature is suitable for forming microemulsion droplets, and that the microemulsion with high water concentration is likely to absorb more surfactants to structure the interface.

8.5.1 MICROEMULSION DETERGENT

Microemulsions are used in many different applications in everyday life (Friberg et al., 2003; Sjoblom, 2001; Friberg, 1976; Birdi, 2016). Liquid detergent formulation is one example. A light-duty microemulsion liquid detergent composition, useful for removing greasy soils from surfaces with both neat and diluted forms of the detergent composition, has been reported. It consists of the following components:

- 1%–10%: a moderately water-soluble complex of anionic and cationic surfactants in which the anionic and cationic moieties are in essentially equivalent or equimolar proportions (an anionic detergent)
- 1%–5%: a cosurfactant
- 1%–5%: an organic solvent
- 70%: water

The recipe is based on the following considerations. It is known that if one
mixes anionic (such as SDS) detergent with a cationic (such as cetyltrimethylam-
monium bromide (CTAB)), then a complex (molar ratio 1:1) is formed, which is
sparingly soluble in water. The reason being that positive and negatively charged
moieties interact and produce a neutral complex (which is insoluble in water). This
complex is oil soluble. The complex component is one in which the anionic and
cationic moieties include hydrophilic portions or substituents, in addition to the
complex-forming portions thereof, the anionic detergent is a mixture of higher
paraffin sulfonate and higher alkyl polyoxyethylene sulfate. The cosurfactant is
a polypropylene glycol ether, a poly-lower alkylene glycol lower alkyl ether, or a
poly-lower alkylene glycol lower alkanoyl ester, and the organic solvent is a nonpolar oil, such as an isoparaffin, or an oil having polar properties, such as a lower
fatty alkyl chain.

This liquid detergent has been reported to be an effective light-duty microemulsion liquid, which is useful for the removal of greasy soils from substrates, both in
neat form and when diluted with water.

8.5.2 MICROEMULSION TECHNOLOGY FOR OIL RESERVOIRS

Enhanced recovery (EOR) is going to be of major interest in the coming decades.
EOR can give rise to increased production from oil reservoirs that are becoming less
productive (Santanna et al., 2009; Bera and Mandal, 2015). The application of surfactant solutions and microemulsions is being investigated. Laboratory experiments
have indicated a rate of over 70% oil recovery by applying microemulsions. They are
effective mainly due to very low IFT at the oil–water interface.

A new microemulsion additive has been developed that is effective in remediat-
ing damaged wells and is highly effective in fluid recovery and relative permeability
enhancement when applied in drilling and stimulation treatments at dilute concentrations (Santanna et al., 2009). The microemulsion is a unique blend of biodegrad-
able solvent, surfactant, cosolvent, and water. The nanometer-sized structures are
modeled with structures, which, when dispersed in the base treating fluid of water
or oil, permit a greater ease of entry into a damaged area of the reservoir or fracture system. The structures maximize surface-energy interaction by expanding up to
12 times their individual surface areas to allow maximum contact efficiency at low
concentrations (0.1%–0.5%). Higher loadings of the order of 2% can be applied in
the removal of water blocks and polymer damage. Laboratory data are shown for the
role of microemulsions in speeding the clean-up of injected fluids in tight gas cores.
Further tests show that the microemulsion additive results in lower pressures to displace fracturing fluids from propped fractures, resulting in lower damage and higher

production rates. This reduced pressure is also evident in pumping operations where friction is lowered by 10%–15% when the microemulsion is added to fracturing fluids. Field examples are shown for remediation and fracture treating of coals, shales, and sandstone reservoirs, where productivity is increased by 20%–50%, depending on the treatment parameters.

References

Adam, N. K., *The Physics and Chemistry of Surfaces*, Clarendon Press, Oxford, 1930.

Adamson, A. W. and Gast, A. P., *Physical Chemistry of Surfaces*, 6th edn, Wiley-Interscience, New York, 1997.

Aderibigbe, A. A., Shale hydraulic fracture, MSc thesis, Texas A&M University, College Station, TX, 2012.

Ahmed, T., *Reservoir Engineering Handbook*, Gulf Professional Publishers, Boston, MA, 2001.

Ahmad, Z., *Principles of Corrosion Engineering*, Elsevier, New York, 2006.

Allan, A. M. and Mavko, G., The effect of adsorption and Knudsen diffusion on the steady-state permeability of microporous rocks, *Geophysics*, 75, 78, 2013.

Aman, Z., In: Gordon Research Conference, Galveston, TX, 28 February, 2016.

Ambrose, R. J., Diaz-Campos, M., Akkutlu, I. Y. and Sondergeld, C. H., SPE-131772, Unconventional Gas Conference, Pittsburgh, PA, 23 February 2010.

Ambrose, R. J., Hartman, R. C. and Akkutlu, I. Y., SPE: Production and Operations Symposium, Oklahoma, OK, SPE-141416-MS, 2011.

Ambrose, R. J., Hartman, R. C., Diaz-Compos, M., Akkutulu, I. Y. and Sondergeld, C. H., Shale gas in calculations Part I: New pore-scale considerations, SPE Journal, 219, 17, 2012.

Attard, P., Electrolytes and electric double layer, *Advances in Chemical Physics*, 92, 1, 1996.

Auroux, A., *Calorimetry and Thermal Methods in Catalysis*, Springer, Berlin, 2013.

Aveyard, R. and Hayden, D. A., *An Introduction to Principles of Surface Chemistry*, Cambridge University Press, London, 1973.

Avnir, D., ed., *The Fractal Approach to Heterogeneous Chemistry*, Wiley, New York, 1989.

Badra, H., Modeling of fractures, MSc Thesis, University of Oklahoma, Norman, OK, 2011.

Bancroft, W. D., *Applied Colloid Chemistry*, McGraw-Hill, New York, 1932.

Bear, J., *Dynamics of Fluids in Porous Media*, Dover, New York, 1972.

Becher, P., *Emulsions, Theory and Practice*, 3rd edn, Oxford University Press, New York, 2001.

Belsky, T., Organic geochemistry and chemical evolution, PhD Thesis, University of California, Berkeley, 1966.

Bera, A. and Mandal, A., Microemulsions: A novel approach to enhanced oil recovery: A review, *Journal of Petroleum Exploration and Production Technology*, 255, 5, 2015.

Bhattacharya, J. and MacEachern, J.A., Models for shelf shales, *JSR*, 79, 184, 2009.

Bihl, J. and Brady, P. V., 125th Anniversary Annual Meeting and Expo. of Geological Society of America, Denver, CO, 27 October, 2013.

Binnig, G. and Rohrer, H., Scanning tunneling microscopy, *Surface Science*, 236, 126, 1983.

Birdi, K. S., *Journal of Colloid and Polymer Science*, 250, 7, 1972.

Birdi, K. S., Cell adhesion on solid surfaces, *J. Theor. Biol.*, 1, 93, 1981.

Birdi, K. S., *Journal of Colloid and Polymer Science*, 260, 8, 1982.

Birdi, K. S., *Lipid and Biopolymer Monolayers at Liquid Interfaces*, Plenum, New York, 1989.

Birdi, K.S., Vu, D. and Winter, A., A study of the evaporation of small water drops placed on a solid, surface, *J. Phys. Chem.*, 93, 3702, 1989.

Birdi, K. S., *Fractals in Chemistry, Geochemistry and Biophysics*, Plenum, New York, 1993.

Birdi, K. S., ed., *Handbook of Surface and Colloid Chemistry*, CRC Press, Boca Raton, FL, 1997.

Birdi, K. S., *Self-Assembly Monolayer (SAM) Structures*, Plenum, New York, 1999.

Birdi, K. S., ed., *Handbook of Surface and Colloid Chemistry-CD Rom*, 2nd edn, CRC Press, Boca Raton, FL, 2003a.

Birdi, K. S., *Scanning Probe Microscopes (SPM)*, CRC Press, Boca Raton, FL, 2003b.

Bird, R. B., *Transport Phenomena*, Wiley, New York, 2007.

Birdi, K. S., ed., *Handbook of Surface and Colloid Chemistry-CD Rom*, 3rd edn, CRC, Boca Raton, FL, 2009.

Birdi, K. S., *Interfacial Electrical Phenomena*, CRC Press, Boca Raton, FL, 2010a.

Birdi, K. S., *Surface and Colloid Chemistry*, CRC Press, Boca Raton, FL, 2010b.

Birdi, K. S., *Surface Chemistry Essentials*, CRC Press, Boca Raton, FL, 2014.

Birdi, K. S., ed., *Handbook of Surface and Colloid Chemistry-CD Rom*, 4th edn, CRC Press, Boca Raton, FL, 2016.

Birdi, K. S. and Ben-Naim, A., Standard free energy of transfer of a solute from water into micelles, *Journal of Chemical Society, Faraday Transactions*, 2035, 76, 1980.

Birdi, K. S., Vu, D. T., Winter, A. and Naargard, A., A study of the evaporation rates of small water drops placed on a solid surface, *Journal of Colloid and Polymer Science*, 266, 5, 1988.

Bissonnette, B., Courard, L. and Garbacz, A., *Concrete Surface Engineering*, CRC Press, Boca Raton, FL, 2015.

Blomberg, C., *Physics of Life*, Elsevier, New York, 2007.

Bockris, J. O., Conway, B. E. and Yeager, E., *Comprehensible Treatise of Electrochemistry*, Vol. 1, Plenum, New York, 1980.

Bolt, P. and Kaldi, J., eds, *Evaluating Fault and Cap Rock Seals*, American Association of Petroleum Geologists, Tulsa, OK, 2005.

Bondarenko, V. Kovalesvka, I. and Ganuschevch, K., *Coal Bed Methane*, CRC Press, Boca Raton, FL, 2014.

Borysenko, A., Clennell, B., Sedev, R., Burgar, I., Ralston, J., Rven, M. and Brandt, A. R., Wettability of clays and shales, *Environmental Science and Technology*, 7489, 42, 2008.

Boschee, P., *Oil and Gas Facilities*, 22, 1, 2012.

Bozak, R. E. and Garcia, M. Jr., Chemistry in the oil shales, *Journal of Chemical Education*, 154, 53, 1976.

Brezonik, P. L. and Arnold, W. A., *Water Chemistry*, Oxford University Press, Oxford, 2011.

Bumb, A. C. and McKee, C. R., Adsorption of methane on solids, *SPE*, 179, 3, 1988.

Burlingame, A. L, Haug, P., Belsky, T. and Calvin, M., Occurance of stearanes in shales, *Proceedings of the National Academy of Sciences*, 1406, 54, 1965.

Burnham, A. K. and McConaghy, J. R., 26th Oil Shale Symposium, Lawrence Livermore National Laboratory, Golden, CO, 2006.

Cahoy, D. R., Gehman, J. and Lei, Z., Fracking patents: The emergence of patents as information-containment tools in shale drilling, *19th Michigan Telecommunications and Technology Law Review*, 279, 2013.

Calvin, M., *Chemical Evolution*, Clarendon, Oxford, 1969.

Cernica, J. N., *Geotechnical Engineering*, Holt, Reinhart and Winston, New York, 1982.

Chalmers, G. R. I. and Busstin, R. M., Surface characterization of coals, *International Journal of Coal and Geology*, 223, 39, 2007.

Chang, Y., Liu, X. and Christie, P., Gas evolution in China, *Environmental Science and Technology*, 12281, 46, 2012.

Chapman, E. C., Capo, R. C., Stewart, B. W., Kirby, C. S., Hammck, R., Schroeder, K. T. and Ederborn, H. M., Isotope characterization of produced water, *Environmental Science and Technology*, 3545, 46, 2012.

Chattoraj, D. K. and Birdi, K. S., *Adsorption and the Gibbs Surface Excess*, Plenum, New York, 1984.

Cini, R., Loglio, G. and Ficalbi, A., Surface tension of water, *Journal of Colloid and Interface Science*, 41, 287, 1972.

Civan, F., *Transport in Porous Media*, 375, 82, 2010.

Clark, D. C., Wilde, P. J. and Marion, D., *Journal of the Institute of Brewing*, 23, 100, 1994.

Clark, R. C., Application of hydraulic fracturing to the stimulation of oil and gas production, *Drilling and Production*, 113, 453, 1953.

Coppens, M. O., Characterization of fractal surface roughness and its influence on diffusion and reaction, *Colloids Surfaces*, 257, 187, 2001.

Corrin, M. L., Areas of carbon blacks and other solids by gas absorption, *Journal of the American Chemical Society*, 4061, 73, 1951.

Cronin, M. T. D., *Predicting Chemical Toxicity and Fate*, CRC Press, New York, 2004.

Davies, J. T. and Rideal, E. K., *Interfacial Phenomena*, Academic, New York, 1963.

Deam, J. R. and Mattox, R. N., Interfacial tension in hydrocarbon systems, *Journal of Chemical and Engineering Data*, 216, 15, 1970.

Defay, R., Prigogine, I., Bellemans, A. and Everett, D. H., *Surface Tension and Adsorption*, Longmans, Green, London, 1966.

De Gennes, P. G., Wyatt, F. B. and Quere, D., *Capillarity and Wetting Phenomena*, Springer, New York, 2003.

Deghanpour, H., Lan, Q., Saeed, Y., Fei, H. and Qi, Z., Spontaneous imbibition of brine and oil in gas shales: Effect of water adsorption and resulting microfractures, *Energy and Fuels*, 3039, 27, 2013.

Dewhurst, D. and Liu, K., *Journal of Geophysical Research: Solid Earth*, 114, 883, 2009.

Dickenson, E., *Colloid Surface, B*, 197, 20, 1992.

Dolphin, D., ed., *The Porphyrins, Structure and Synthesis*, Academic, New York, 1978.

Donaldson, E., Alam, W. and Begum, N., *Hydraulic Fracturing Explained*, Gulf, Houston, TX, 2013.

Donaldson, M. A., Berke, A.E. and Raff, J. D., Uptake of gas on soil surfaces, *Environmental Sciences and Technology*, 48, 375, 2013.

Drummond, C. and Israelachvili, J., *Journal of Petroleum Science and Engineering*, 61, 45, 2004.

Dunning, J.D., Wardell, D. and Dunn, D.E., Chemo-mechanical weakening in the presence of surfactants, *J. Geophysical Research (Solid Earth)*, 85, 5344, 1980.

Dyni, J. R., *Oil Shale*, 193, 20, 2003.

El-Shall, H. and Somasundaran, P., Fracture formation, *Powder Technology*, 275, 38, 1984.

Emmett, P. H. and Brunauer, S., *Journal of the American Chemical Society*, 1553, 59, 1937.

Engelder, T., Cathles, L. M. and Bryndzia, L. T., Residual water treatment in gas shales, *Journal of Unconventional Oil and Gas Resources*, 33, 7, 2014.

Fainerman, V. B., Miller, R. and Mohwald, H., *Journal of Physical Chemistry*, 809, 106, 2002.

Fathi, E. and Yucel, A. I., SPE Annual Technical Conference and Exhibition, New Orleans, LA, 4, 2009.

Feder, J., *Fractals: Physics of Solids and Liquids*, Plenum, New York, 1988.

Fendler, J. H. and Fendler, E. J., *Catalysis in Micellar and Macromolecular Systems*, Academic, New York, 1975.

Fengpeng, L., Zhiping, L., Zhifeng, L., Zhihao, Y. and Yingkun, F., Oil and gas technology, *Review IFP Energies Nouvelles*, 1191, 69, 2014.

Feres, R. and Yablonsky, G., Knudsen's cosine law and random billiards, *Chemical Engineering Science*, 1541, 59, 2004.

Freeman, C. M., Moridis, G. J. and Blasingame, T. A. A., *Proceedings: TOUGH Symposium*, Berkeley, 14 September 2009.

Freeman, C. M., Moridis, G. J. and Blasingame, T. A. A., *Transport in Porous Media*, 90, 253, 2011.

Freundlich, H., *Colloid and Capillary Chemistry*, Methuen, London, 1926.

Friberg, S., Larsson, K. and Sjoblom, J., *Food Emulsions*, CRC Press, Boca Raton, FL, 2003.

Frohn, A. and Roth, N., *Dynamics of Droplets*, Springer, Berlin, 2000.

Fuerstenau, M. C., Jameson, G. J. and Yoon, R. H., *Froth Flotation*, Society of Mining Metallurgy and Exploration, CO, 1985.

Gaines, G. L., Jr., *Insoluble Monolayers at Liquid-Gas Interfaces*, Wiley-Interscience, New York, 1966.

Gitis, N. and Sivamani, R, *Tribology Transactions*, Taylor & Francis Group, New York, 2014.

Gold, T. and Soter, S., The deep-earth gas hypothesis, *Scientific American*, 154, 242, 1980.

Gregory, K. B., Vidic, R. D. and Dzombak, D. A., Water management challenges associated with the production of shale gas by hydraulic fracturing, *Elements*, 181, 7, 2011.

Gudmundsson, A., Fracture dimensions, displacements and fluid transport, *Journal of Structural Geology*, 1221, 22, 2000.

Gupta, D. V. S. and Carman, P. S., Paper SPE-141260, at SPE International Symposium on Oilfield Chemistry, The Woodlands, TX, 11 April 2011.

Gupta, D. V. S. and Hlidek, B. T., Paper SPE-119478, at SPE Hydraulic Fracturing Tech. Conf., The Woodlands, TX, 21 January 2009.

Hansch, C., Leo, A. and Taft, W., *Chemical Reviews*, 783, 102, 2002.

Hansen, M. C., *Hansen Solubility Parameters*, Taylor & Francis, New York, 2007.

Harkins, W. D., *The Physical Chemistry of Surface Films*, Reinhold, New York, 1952.

Hesselbo, S., Grocke, D., Jenkyns, H. C., Bjerrum, C. J, Farrimond, P., Bell, H. S. M. and Green, O. R., *Nature*, 392, 406, 2000.

Hill, D.G. and Nelson, C.R., Gas productive fractured shales, *Gas Tips*, 6, 4, 2000.

Holmberg, K., ed., *Handbook of Applied Surface and Colloid Chemistry*, Wiley, New York, 2002.

Howard, G. C., *Hydraulic Fracturing*, Mineral Law Series, Rocky Mountain Mineral Law Foundation, Westminster, CO, 1970.

Howarth, R. W., Santoro, R. and Ingraffea, A., *Climate Change*, 679, 106, 2011.

Hsu, J. P. and Kuo, Y. C., The critical coagulation concentration of counterions: Spherical particles in asymmetric electrolyte solutions, *Journal of Colloid and Interface Science*, 530, 185, 1997.

Hunt, J., *Petroleum Geochemistry*, W.H. Freeman, San Francisco, CA, USA, 1996.

Hu, K. and Bard, A. J., In situ monitoring of kinetics of charged thiol adsorption on gold using an atomic force microscope, *Langmuir*, 14, 4790, 1998.

Ivanov, I. B., Effect of surface mobility on the dynamic behavior of thin liquid films, *Pure and Applied Chemistry*, 1241, 52, 1988.

Ivanov, I. B. and Kralchevsky, P., *Colloid Surfaces A*, 155, 128, 1997.

Jarvie, D. M., Hill, R. J., Ruble, T. E., and Pollastro, Unconventional shale-gas systems, *American Association of Petroleum Geologists Bulletin*, 475, 91, 2007.

Javadpour, F., Gas flow in shales, *Journal of Canadian Petroleum Technology*, 16, 16, 2009.

Javadpour, F., Farshi, M. M. and Amrein, M., AFM studies of shale, *Journal of Petroleum Technology*, 236, 2012.

Javadpour, F., Fisher, D. and Usworth, M., Nanopores in shales and siltstones, *Journal of Petroleum Technology*, 16, 46, 2007.

Jaycock, M. J. and Parfitt, G. D., *Chemistry of Interfaces*, Ellis Horwood Hemel Hempstead, England, 1981.

Jennings, H. Y., The effect of temperature and pressure on the interfacial tension of benzene-water and normal decane-water, *Journal of Colloid Interface Science*, 323, 24, 1967.

Josh, M., Estaban, L., Delle, C., Sarout, J., Dewhurst, D. N. and Clennel, M. B., *Journal of Petroleum Science and Engineering*, 107, 88, 2012.

Kale, S. V., Rai, C. S. and Sondergeld, C. H., SPE-131770, Unconventional Gas Conference, Pittsburgh, PA, 3 February 2010.

Kamperman, M. and Synytska, A., *Journal of Material Chemistry*, 22, 19390, 2012.

Kargbo, D. M., Wilhelm, R. G. and Campbell, D. J., *Environmental Science and Technology*, 5679, 44, 2010.

Kim, J. H., Ahn, S. J. and Zin, W. C., *Langmuir*, 6163, 23, 2007.

Klimpel, R. R., in: Kawatra, S. K., ed., *High Efficiency Coal Preparation*, Society for Mining, Metallurgy and Exploration, Littleton, CO, 1995.

Klimpel, R. R., Society of Mining, Metallurgy, and Exploration, Littleton, 141, 1995.

Knudsen, M., Kinetic Theory of gases, *Methuens Monographs on Physical Subjects*, Methuen Publishing, London, 1952.

Kortum, G., *Treatise on Electrochemistry*, Elsevier, New York, 1965.

Kubinyi, H., *QSAR*, VCH, Weinheim, 1993.

Kumar, A., MSc thesis, Adsorption of Methane on Carbon, National Institute of Technology, Rourkela, India.

Kumar, D., *Analysis of Multicomponent Seismic Data (Offshore Oregon)*, Ph.D., The University of Texas, Austin, TX, 2005.

Kvenvolden, K. A., A review of the geochemistry of methane in natural gas hydrate, *Organic Geochemistry*, 997, 23, 1995.

La Mer, V. K., *Retardation of Evaporation by Monolayers*, Academic, New York, 1962.

Lash, G. G., Analyses of black shales, *Marine and Petroleum Geology*, 317, 23, 2006.

Lash, G. G. and Blood, D. R., Analyses of black shales, *Basin Research*, 51, 19, 2007.

Lash, G. G. and Engelder, T., Analyses of black shales, *American Association of Petroleum Geologists Bulletin*, 1433, 89, 2005.

Lash, G. G. and Engelder, T., Analyses of black shales, *American Association of Petroleum Geologists Bulletin*, 61, 95, 2011.

Latanision, R. M. and Pickens, J. R., *Atomistics of Fracture*, Plenum, New York, 1983.

Leja, J., *Surface Chemistry of Froth Flotation*, Springer Science and Business Media, New York, 2012.

Letham, E. A., Matrix permeability measurement of gas shales, Thesis, The University of British Columbia, Canada, 2011.

Levorsen, A. I., *Geology of Petroleum*, 2nd edn, Freeman, San Francisco, 1967.

Lichtman, V. I., Rehbinder, P. A. and Karpenko, G. V., *Effect of Surface-Active Medium on the Deformation of Metals*, H. M. Stationery Office, London, 1958.

Liebowitz, H., ed., *Fracture: An Advanced Treatise*, Academic, New York, 1971.

Liu, Y., Yu, B., Xu, P. and Wu, J., *Fractals*, 1, 55, 2007.

Lovett, D., *Science with Soap Films*, Institute of Physics Publishing, Bristol, 1994.

Lu, X. C., Li, F. C. and Watson, A. T., *Fuel*, 590, 74, 1995.

Ma, Y. Z. and Holditch, S. A., *Unconventional Oil and Gas Resources Handbook*, Elsevier, Amsterdam, 2016.

McCafferty, E., *Introduction to Corrosion Science*, Springer, New York, 2010.

Malkin, A. I., *Colloid Journal*, 74, 223, 2012.

Matijevic, E., ed., *Surface and Colloid Science*, Vol. 1–9, Wiley-Interscience, New York, 1969–1976.

Maynard, J. B., Geology, *GSA*, 262, 9, 1983.

Melikoglu, M., Gas shale reservoirs, *Renewable and Sustainable Energy Reviews*, 460, 37, 2014.

Mirchi, V., Saraji, S., Goual, L. and Piri, M., *Fuel*, 148, 127, 2015.

Morrow, N. R. and Mason, G., *Current Opinion in Colloid and Interface Science*, 6, 321, 2001.

Nduagu, I. and Gates, I. D., *Environmental Science and Technology*, 8824, 49, 2015.

O'Brien, N. and Slatt, R. M., eds, *Argilleaceous Rock Atlas*, Springer, New York, 1990.

Oligney, R. and Economides, M., *Unified Fracture Design*, Orsa, 2002.

Ozkan, E., Performance of horizontal wells, Thesis, Tulsa University, OK, 1988.

Ozkan, E., Raghavan, R. and Apaydin, O. G., SPE, Annual Technical Conference and Exhibition, Florence, Italy, 19, 2010.

Pagels, M., Hinkel, J. and Willberg, D., *Capillary Suction*, SPE, Inter. Symp. and Exh. On Formation and damage control, SPE-151832, Denver, CO, 2011.

Partington, J. R., *An Advanced Treatise of Physical Chemistry*, Vol. II, Longmans Green, New York, 1951.

Passey, Q., Bohacs, K., Esch, W., Klimintidis, R. and Sinha, S., *Shale Reservoir Model*, SPE-131350, International Oil and Gas Conference and Exhibition, Beijing, China, 2010.

Perenchio, W. F., Corrosion of reinforcing steel, *ASTM STP*, 169C, 164, 1994.

Rao, V., Shale Gas, RTI Press, Research Triangle Park, NC, 2012.

Rehbinder, P. A. and Schukin, E. D., *Progress in Surface Science*, 3, 97, 1972.

Reid, R. C., Prausnitz, J. M. and Poling, B. E., *The Properties of Gases and Liquids*, 4th edn, McGraw-Hill, New York, 1987.

Richard, F. S. and Qin, B., *Petrophysics*, 49, 301, 2008.

Roberge, P. R., *Handbook of Corrosion Engineering*, McGraw-Hill, New York, 1999.

Rogala, A., Krzysiek, J., Bernaciak, M. and Hupka, J., *Physiochemical Problems of Mineral Processing*, 313, 49, 2013.

Rosen, J. and Kunjappu, J. T., *Surfactants and Interfacial Phenomena*, Wiley, New York, 2012.

Ross, D. J. K. and Bustin, R. M., *Fuel*, 2696, 86, 2007.

Rubinstein, B. Y. and Bankoff, S. G., *Langmuir*, 17, 130, 2001.

Russell, W. I., *Principles of Petroleum Geology*, McGraw-Hill, New York, 1960.

Ruthven, D. M, *Principles of Adsorption and Adsorption Processes*, Wiley-Interscience, New York, 1984.

Sakhaee-Pour, A. and Bryant, S., *SPE- Reservoir Evaluation and Engineering*, 15, 401, 2012.

Santanna, V. C., Dantas, C. and Neto, A. A. D., *Journal of Petroleum Science and Engineering*, 117, 66, 2009.

Santos, P.M., *Automatic Pavement Crack Detection*, M.Sc., Instituto Superior Tecnico, Univ. Tech. Lisbon, Portugal, 2008.

Scesi, L. and Gattinoni, *Water Circulation in Rocks*, Springer, New York, 2009.

Scheidegger, A. E., *The Physics of Flow through Porous Media*, University Press of Toronto, Toronto, 1957.

Scheider, M., Osselin, F., Andrews, B., Rezgui, F. and Tebeling, P., *Journal of Petroleum Science and Engineering*, 476, 78, 2011.

Scheludko, A., *Colloid Chemistry*, Elsevier, New York, 1966.

Schoell, M., *Geochemica et Cosmochimica Acta*, 649, 44, 1980.

Schramm, L. L., *Surfactants: Fundamentals and Applications in Industry*, Cambridge University Press, Cambridge, 2010.

Shabro, V., Javadpour, F. and Torres-Verdí, C., A generalized finte-difference diffusive-advective (FDDA) model for gas flow in micro- and nano- porous media, *World Journal of Engineering*, 7, 6, 2009.

Shabro, V., Torres-Verdí, C. and Javadpour, F., Society of Petrophysicists and Well-Log Analysts, Colorado Springs, CO, 15 May, 2011a.

Shabro, V., Torres-Verdí, C., and Javadpour, F., SPE-144355, paper presented at the Unconventional Gas Conference, SPE, The Woodlands, TX, 14 June, 2011b.

Shabro, V., Torres-Verdín, C. and Sepehrnoori, K., SPE, Tech Conference, Austin, TX, October, 2012.

Shabro, V., Modeling of Organic shale, PhD Thesis, U. Texas, Austin, TX, 2013.

Shih, J. S., Saiers, J. E., Anisfeld, S. C. and Climstead, S., *Environment Science and Technology*, 9557, 49, 2015.

Shipilov, S. A., Jones, R. H., Olive, J. M. and Rebak, R. B., eds, *Chemistry, Mechanics and Mechanisms*, Elsevier, New York, 2008.

Shou, D., Ye, L., Fan, J. and Fu, K., *Langmuir*, 149, 30, 2014.

Singh, P., Stratigraphic of shale, Northeast Texas, PhD Thesis, University of Oklahoma, Norman, OK, 2008.

Slatt, R., *Open Geosciences*, 135, 3, 2011.

Sloan, E. D. and Koh, C., *Clathrate Hydrates of Natural Gases*, 3rd edn, CRC Press, Boca Raton, FL, 2007.

Smith, M. B. and Montgomery, C. T., *Hydraulic Fracturing*, CRC Press, Boca Raton, FL, 2015.

Soltis, A. N., Chen, J., Atkin, L. Q. and Hendy, S., Specific ion binding influences on surface potential of chromium oxide, *Current Applied Physics*, 4, 152, 2004.

Somasundaran, P., *Colloidal and Surfactant Sciences*, CRC Press, New York, 2006.

Somasundaran, P., *Oil Spill Remediation*, Wiley, New York, 2010.

Somasundaran, P., *Encyclopedia of Surface and Colloid Science*, CRC Press, Boca Raton, FL, 2015.

Somasundaran, S., Freeman, M. P. and Fitzpatrick, J. A., eds, *Theory, Practice and Process for Separation*, Engineering Foundation, New York, 1981.

Somorjai, G. A., *Introduction to Surface Chemistry and Catalysis*, Wiley, New York, 2000.

Sondergeld, C. H., Ambrose, R. J., Rai, C. S. and Moncrieff, J., Unconventional Gas Conference, SPE-131771-PP, Pittsburgh, PA, 23 February, 2010.

Speight, J. G., ed., *Fuel Science Technology Handbook*, Marcel Dekker, New York, 1990.

Speight, J. G., *The Chemistry and Technology of Coal*, 3rd edn, CRC Press, Boca Raton, FL, 2013.

Starostina, I. A., Stoyanov, O. V. and Deberdeev, R. Y., *Polymer Surfaces: Adhesion on Polymer-Metal Systems*, CRC, Boca Raton, FL, 2014.

Stefan, J., Surface tension at oil/water interface, *Ann. Phys.*, 29, 655, 1886.

Stegemeier, G. I., PhD Thesis, University of Texas, Austin, TX, 1959.

Stringfellow, W. T. and Domen, J. K., *Journal of Hazardous Materials*, 37, 275, 2014.

Striolo, A., Klaessig, F., Cole, R., Wilcox, J., Chase, G. G., Sondergeld, C. H. and Pasquali, M., Workshop Report, NSF, 2012.

Tanford, C., *The Hydrophobic Effect*, Wiley, New York, 1980.

Theodori, G.L., Luloff, A.E., Willits, F.K. and Burnett, D.B., Hydraulic fracturing, *Energy Res. and Soc. Sci.*, 2, 66, 2014.

Thomas, M. M. and Clouse, J. A., *Geochimica et Cosmochimica Acta*, 2781, 54, 1990.

Tissot, B. and Welte, D. H., *Petroleum Formation and Occurrence*, Springer, New York, 1978.

Tissot, B. P. and Welte, D. H., *Petroleum Formation and Occurrence*, Springer, Berlin, 1984.

Trevena, D. H., *The Liquid Phase*, Wykeham Science Series, London, 1975.

Tucker, M. E., *Sedimentary Petrology, an Introduction*, Blackwell, London, 1988.

Tunio, S. Q., Tunio, A. H., Ghirano, N. A. and Adawy, Z. M. E., *International Journal of Applied Science and Technology*, 143, 1, 2011.

Vafai, K., *Handbook of Porous Media*, CRC Press, Boca Raton, FL, 2015

Venkateswarlu, K. S., *Water Chemistry*, New Age International Publishers, New Delhi, 1996.

Walls, J., *Journal of Petroleum Technology*, 2708, 34, 1982.

Wang, Q., Chen, X., Jha, A. N. and Rogers, H., *Renewable and Sustainable Energy Reviews*, 1, 30, 2014.

Wang, S., Feng, Q., Javadpour, F., Xia, T. and Li, Z., *International Journal of Coal Geology*, 147, 9, 2015.

Weber, C. L. and Clavin, C., *Environmental Science and Technology*, 5688, 46, 2012.

Wilson, A. J., ed., *Foams: Physics, Chemistry and Structure*, Springer, New York, 1989.

Wu, X. Z., Ocko, B. M., Sirota, C. B., Sinha, S. K., Deutsch, M., Cao, B. H. and Kim, M. W., X-ray scattering of liquid surfaces, *Science*, 1018, 261, 1993.

Yen, T. F. and Chilingarian, G. V., eds, *Oil Shale*, Elsevier, Amsterdam, 1976.

Yew, C. H. and Weng, X., eds, *Mechanics of Hydraulic Fracturing*, 2nd edn, Elsevier, Amsterdam, 2014.

Yoon, R. H., Luttrell, G. H. and Adel, G. T., *Advanced System for Producing Superclean Coal*, Final Report, DOE, August, 1990.

Yu, B. and Li, J., *Fractals*, 365, 9, 2001.

Yu, H. Z., Soolaman, D. M., Rowe, A.W. and Banks, J. T., *ChemPhysChem*, 5, 1035, 2004.

Yu, Y. S., Wang, Z. and Zhao, Y. P., *Journal of Colloid and Interface Science*, 1016, 10, 2011.

Zelenev, A.S., *Surface Energy of Shales*, Soc. Petr. Eng., The Woodlands, TX, 11–13 April, 2011.

Zhang, K., Fluids for fracturing subterranean formations, US Patent, No.6,468,945, 2002.

Zhang, K. and Gupta, D.V.S., Foam fluid for fracturing, US Patent No.6,410,489, 2002.

Zhang, Y., Chen, Y., Shi, L. and Guo, Z., Stable superhydrophobic coatings from thiol-ligand nanocrystals and their application in oil/water separation *Journal of Material Chemistry*, 22, 799, 2012.

Zheng, M., MS Thesis, Rock-based characterization of gas shale, University of Oklahoma, Norman, OK, 2011.

Zinola, C. F., *Electrocatalysis: Computational, Experimental and Industrial Aspects, Surfactant Science Series*, CRC Press, Taylor & Francis Group, Boca Raton, FL, 2010.

Zou, C., *Unconventional Petroleum Geology*, Elsevier, Amsterdam, 2012.

Appendix I: Geochemistry of Shale Gas Reservoirs (Shale and Energy)

Since man discovered *fire, energy* has been an important everyday essential part of life on earth. Initially, the energy need was provided by wood and later other sources were added (such as coal, oil, gas, wind, hydro energy, atomic energy, sun energy, wave, etc.). Further, the interior of the earth is known to be at a very high potential as compared to the surface of the earth (regarding temperature and pressure gradient) (Dewhurst and Liu, 2009; Hunt, 1979; Gold and Soter, 1980). Hence, there is a very dynamic difference between the interior and the surface of the earth. This indicates that the interaction between mankind (and evolution) and the changes in the interior of the earth are interrelated. Mankind is aware of this from the lava eruptions witnessed from various active sites. Further, this is well known from the fact that about a century ago, oil was discovered when wells were drilled for drinking water (in the United States and elsewhere).

Man on earth needs energy to survive. For example, oil alone is consumed at a rate of *~100 million barrels per day*! A decade ago, the supplies of oil and gas, as found in conventional reservoirs, were running low (especially in the North Sea and the United States) (Dewhurst and Liu, 2009). These reservoirs are considered as rock structures where material (oil or gas) had been trapped while it was moving from source reservoirs (where oil or gas was originally created through geological history). Most of the conventional rocks are sandstone or carbonate material. Exploration and development of these inaccessible sources of energy as found in shale deposits (worldwide) are accepted to have potential regarding the future global energy demand. The rock structure in which the oil/gas originally formed is designated as source rock.

(Oil/gas migration):

Source rock (shale) → conventional reservoir

It has been known for many decades that the oil/gas as found in conventional reservoirs had migrated from source rock (i.e., shale reservoirs [nonconventional]). Further, oil/gas deposits in conventional reservoirs are found in sedimentary rocks. Oil pools are found in underground structures where this fluid has been trapped in the sedimentary rock layers, which are folded. These oil/gas reservoirs are porous sedimentary rock formation, capped with a layer of impermeable structure formation (through which oil/gas cannot migrate easily). It was known from these early explorations that the oil/gas in conventional reservoirs was not created there (Levorsen, 1967; Hunt, 1979; Maynard, 1983; Tissot and Welte, 1984; Russell, 1960; Speight, 1990; Dyni, 2003; Slatt, 2011; Melikoglu, 2014). It is also well accepted that in these reservoirs, the oil has migrated through the layers until trapped. Currently, it is

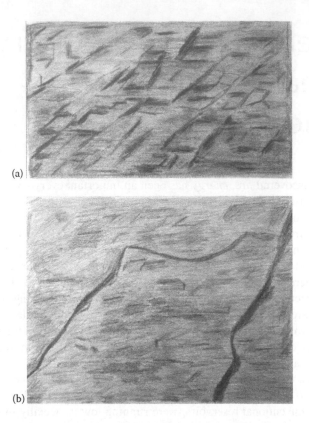

FIGURE A1.1   A typical drawing of shale rock (a); fracture in a shale rock (micrometer scale) (b).

established that shale reservoir is the source rock of oil and gas (mostly methane). The typical layer-cake structure is vividly the most characteristic shale rock (Figure A1.1a and b).

Thus, gas shale is a source reservoir where gas is tightly bound and there has been very little desorption (i.e., a very minute percent of original oil/gas in place has migrated). Furthermore, carbon isotope analyses (over 200 different shale samples) has indicated that the organic material is different in different rock. It was concluded that the nonmarine carbon has an isotopic composition similar to that of coal. The latter suggests that the source was woody material. The study of petroleum geochemistry is based on the understanding of basic chemical–physical principles regarding the origin, creation, and migration of petroleum (including methane, ethane, etc.).

The chemical evolution and geochemistry of earth are known to be complex phenomena (Shabro, 2013; Ozkan, 1988; Hunt, 1979; Calvin, 1969; Lash, 2006; Lash and Blood, 2007; Lash and Engelder, 2005, 2011). The definition of shale is generally used for all kinds of clay-rich sedimentary rocks. Common sedimentary rocks are sandstone, limestone, and shale. Shale is known to be a fine-grained sedimentary material, consisting mostly of flakes of various clay minerals, quartz, calcite, and organic material. Shale has a typical laminate structure. It also has breaks along the thin laminated parallel layers. This kind of rock exhibits low porosity and low permeability. Due to the organic material, black shale is related to carbonaceous content. There is about 60% shale in sedimentary rocks. The mineralogical composition of an average shale is

- Quartz: 30%
- Feldspar: 4%
- Carbonates: 5%
- Iron oxides: 1%
- Clay minerals: 50%
- Organic matter: variable

The so-called black shale is found to yield oil/gas. Natural gas has formed through a thermogenic process of carbonaceous organic material over a long period. The petroleum found in these rocks is designated as oil and natural gas products (mainly consisting of carbon and hydrogen). These are known to have generated from the decomposition of plants or animal remains. One finds that typically, shale formations with gas are around 2–3 km deep in the earth. Analyses of shale rocks are used to determine the recovery strategy of a site (play). It is also found that 2% total organic carbon (TOC) is about the minimum for viable oil/gas recovery. TOC is indicative of the quantity of organic matter that may have decomposed to oil/gas; hence, this is found to be directly proportional to the gas yield/recovery. At one site, the following relationship was found between gas yield and TOC:

$$\text{Gas yield} = 66(\text{TOC}) + 60$$

Obviously, this correlation cannot be expected to be generalizable from site to site. However, it gives a good indication that the TOC is the determining factor as regards oil/gas recovery. It is also found that TOC is probably converted to oil/gas at about 140°C in the reservoir. The oil–clay (illite) adhesion properties have been studied (Bihl and Brady, 2013). This data was found useful in the understanding of hydraulic fracturing and the degree of flow-back. The oil adhesion was found to play an important role in these surface properties of shale. The electrostatic interactions in oil (charged) and calcite (charges) were investigated.

The composition and physical properties of earth vary a great deal as one moves from its interior core to the surface. The core of the earth is at very high temperature (6000°C) and high pressure. Many of the processes found on the surface of the earth cannot exist in the interior of the earth. Another vastly different phenomenon is the existence of plants and living species on the surface of the earth. The plants are created by the combination of CO_2 (from air), water, and photosynthesis (from sunlight):

$$CO_2 + \text{water} + (\text{sunlight}) \rightarrow \text{plants} \qquad (\text{I.1})$$

The reaction in Equation A.1 is mainly possible at temperatures around 25°C, as found close to the equator (Calvin, 1969). Furthermore, all living species on the earth are dependent on water + food (plants) + oxygen (found in the air at ca. 20%).

These observations thus show that both oxygen and CO_2 in air are the most essential gases for the existence of living species on the earth. Currently, one finds that the concentrations of both these gases are relatively stable. The concentration of CO_2 has been increasing during the last few decades (from 300 to 400 ppm). Air (composed of nitrogen [ca. 80%], oxygen [ca. 20%], and carbon dioxide [ca. 400 ppm 0.04%]) is thus the most essential source of different life-sustaining elements. Further, nitrogen in air is used to synthesize ammonia as fertilizer, which is an integral component of plants. Nitrogen as found in air is also a substance, similarly to oxygen and CO_2, that is in equilibrium with different phases in the earth's biocycle.

It is also well established that hydrocarbons (oil and natural gas) are of organic origin (Calvin, 1969; Hunt, 1979; Speight, 1990; Schoell, 1980). It is also assumed from various analyses that shale gas has been generated by various reactions:

- Thermogenic slow degradation of different organic matter (TOC)
- Cracking of oil-like matter
- Biogenic processes which degraded the organic matter

In all cases, it is carbon from CO_2 (present in air at very low concentrations) that has been transformed into hydrocarbons. This was found from extensive analyses of oil, in which traces of molecules found in plants and so on were analyzed. However, some literature reports also argue that gas (methane) may have been formed deep within the earth. As regards oil, of course, the structure itself suggests that it cannot withstand the very high temperatures found inside the core of the earth. Isotopic analyses of carbon have also been used to determine the origin of petroleum molecules. One therefore finds a large amount of literature showing that most of the petroleum originates from plants and animals, which were buried and fossilized in sedimentary rocks (Calvin, 1969; Tissot and Welte, 1984) (Figure A1.1). Over a length of time, petroleum was formed, which later transformed into oil and gas. It is also found that some molecular indicators found in oil, such as porphyrins (Dolphin, 1978), originate from plants and animals. Porphyrins are organic molecules that are similar in structure to both chlorophyll (as found in plants) and hemoglobin (as found in animal blood) (Calvin, 1969; Hunt, 1979; Tissot and Welte, 1984; Levorsen, 1967). It has been shown that on oxidation and heat treatment, the porphyrin molecule is converted to oil and gas. Further, this is supported by the fact that trace amounts of porphyrin molecules are found in oil reservoirs. The breakdown of the latter process is considered to be complex, but some mechanisms have been suggested. Analyses show that the concentration of porphyrin in oil varies from trace amounts to 400 ppm (Calvin, 1969; Hunt, 1979). Another observation is that one finds very large reserves of methane in the form of water hydrates around the surface of the earth (Appendix I). In fact, these gas-hydrate reservoirs are considered potentially important for future gas recovery.

Different shale reservoirs investigated so far are known to be of varying geological age. Analyses have shown that black shales are rich in organic material (ca. 12% TOC) (Calvin, 1969; Engelder et al., 2014; Lash and Engelder, 2011; Slatt, 2011; Badra, 2011).

The deposition of shale took place about 390 million years ago. Rivers carrying clay and silt created the layer-like structures. It is estimated that the organic-rich shales became compact structures (low permeability) under the high pressure (at a depth of a few kilometers). During an estimated 30 million years, under high temperature (ca. 90°C), degradation of the organic material (kerogen) to smaller organic molecules took place. The latter molecules were mainly alkanes and small amounts of other organic molecules (oil) and methane (gas). It is also considered that kerogen originates from plant decomposition. The shale, which may contain organic matter, is thus considered as the source rock (i.e., where petroleum was generated). In some natural processes, where there were fractures and an excess of oil/gas, these substances migrated upward. This migration stopped in the upper structures, which are termed *conventional* reservoirs. It is thus seen that one would find less adsorbed petroleum in conventional reservoirs than in the source rock. Especially, gas recovery from conventional reservoirs will be much greater than in the shale (source rock). The production rates support this conclusion. The unconventional shale gas reservoirs are recognized as being complex structures with a predominantly layered, heterogeneous composition. This structure is also found to give high variability in parallel and perpendicular orientation. This also means that fracturing will be variable in different shale reservoirs (as determined by the composition). The gas production in a shale matrix is dependent mainly on the pore size of the reservoir. These can be considered as being inorganic or organic pore matrices. The diffusion mechanisms in these two different kinds of matrix will be different.

It is also important to mention that the worldwide energy needs are not only essential for mankind, but are also very large. One source alone, oil (ca. 30% of total energy supply), is used at a rate of about 100 million barrels per day. Additionally, it creates a vast demand for transportation worldwide. In this context, one finds that there are about 700 gigatons of world resources of oil shale (5 trillion barrels [760 billion cubic meters] of shale oil) (Fawzi et al., 2014). The biggest currently known shale gas/oil reserves are in the United States, China, and Canada. In China (Chang et al., 2012), shale gas reserves are estimated at 25 trillion cubic meters, which is nearly 200 times China's yearly consumption. It is estimated that in the United States, there are over 20,000 trillion cubic feet of recoverable gas reserves (which could last for over 100 years). Furthermore, there are also questions relating to the *origin* of petroleum (i.e., oil or gas), which are also relevant to the search for more energy supplies (Belsky, 1966; Calvin, 1969; Howarth et al., 2011; Gregory et al., 2011; Chang et al., 2012; Wang et al., 2014; Weber and Clavin, 2012; Fawzi et al., 2014).

In general, the production rates of gas from shale reservoirs are lower than from conventional reservoirs. However, while the gas-recovery rates are lower from shale deposits, these reservoirs are comparatively very large and are expected to produce for a century or more. The amount of original gas-in-place (GIP) for any shale gas reservoir is related to a number of parameters (Richard and Qin, 2008; Kargbo et al., 2010; Freeman et al., 2011; Sakhaee-Pour and Bryant, 2012):

- Size of the shale reservoir
- TOC
- Porosity of the rock
- Extent of water

The amount of gas that can be recovered as compared with total GIP is dependent on

- The porosity (and fracture characteristics)
- Adsorption–desorption equilibria
- The degree of fracture formation

Furthermore, shale rock is composed of organic matter (kerogen, etc.) and inorganic matter. Kerogen is known to behave differently from other shale constituents, because it exhibits hydrocarbon-wetting properties. It is also abundant in nanopores (porous matrix) and capable of adsorbing gas (methane is an organic molecule). On the other hand, it is found that the inorganic matter (feldspar/quartz) is water wet, with low porosity. The latter characteristic means that in the inorganic regions, capillary pressure will be the dominant driving mechanism for water flow, especially during hydraulic fracturing operations. This is found to have major consequences for the degree of fluid recovery (flow-back process). The gas shale recovery is thus composed of different matrices:

- Darcy flow
- Gas diffusion
- Desorption
- Capillary pressure
- Degree of wetting by the hydraulic fluid

It is found that the shale gas reservoirs have complex and variable structures. The organic material (marine or plants), which was buried ages ago, has evolved after high pressure and temperatures (Burligame et al., 1965; Calvin, 1969; Tissot and Welte, 1978). The organic-rich material (TOC) can be as thick as 100 m or more. These rocks are generally analyzed by

- TOC
- X-ray diffraction
- Adsorbed gas indicator
- Core analyses
- Porosity
- Permeability
- Fluid saturation (wetting)
- Optical and electron microscopy
- Density
- Borehole image analysis
- Degree of water wetting

A few decades ago, the invention and application of the novel technology of horizontal drilling and multistage hydraulic fracturing initiated an oil/gas recovery at low cost from low-permeability rock (from 10^{-9} to 10^{-6} D) (Jarvie et al., 2007; Javadpour, 2009; Shabro, 2013; Yu et al., 2004; Wang et al., 2015).

The application of horizontal drilling in oil reservoirs began in the 1980s, and this technology has developed ever since. Especially, major advances have been made in the downhole equipment, which allows better control of oil/gas recovery. The shale reservoirs are known to be very large, and the thickness may vary around 700 m. For example, this technology is known to have increased gas production from 0.4 trillion cubic feet in 2000 to 5 trillion cubic feet in 2010 (in the United States). Shale gas reserves are thought to be found around the world. Some major areas are shown in the following table.

Country	Recoverable Shale Gas (Trillions of Cubic Feet)
Australia	400
Canada	400
Argentina	700
China	1300
France	200
Mexico	700
Poland	200
South Africa	500
United States	900
Total world	6600

This discovery of potential oil/gas recovery from shales has initiated important geological research. Especially, the geomechanical features of shales are recognized as important parameters for fracking. The low porosity (nanoscale) has also led to increased research in such porous materials. The recovery of oil/gas from shale is already a game changer as regards the world economy and geopolitics. In fact, the shale gas technology (mainly in the United States) and its impact on the world energy market is a major twenty-first-century revolution. It has been one of the main factors for a gas price reduction from about 9 U.S. dollars/BTU to ca. 3 U.S. dollars/BTU in 2015. Additionally, it has been the reason for the reduction in the price of other energy sources, such as coal (which is the main energy source in many developing countries [China, India, Africa, etc.]). It is also expected that other countries (such as China, Poland, and Australia) will explore shale gas/oil energy production in the near future. It has been established that the macroscopic and microscopic scale details of shale reservoirs are needed. Seismic imaging provides information at resolutions of tens of meters. Porosity and gas content are considered to be important in shale gas reservoirs. The pore structure and size have been analyzed by using fractal procedures (Feder, 1988; Birdi, 1993; Yu and Li, 2001; Liu et al., 2007). A more detailed analysis is needed to understand fracture formation and gas recovery (Javadpour et al., 2012). The gas-recovery mechanism is related to the adsorption–desorption surface equilibrium. The gas is mostly adsorbed to the surface of the reservoir rock under high pressure. As the fracture is created and the pressure is decreased, desorption will take place (as required by the equilibrium constant).

The interaction of CO_2 in air with water molecules and sun energy (photosynthesis) leads to the formation of plants of all kinds. Not only that, but the earth itself is full of energy sources (coal, oil, and gas). Especially, gas (mostly methane) is present in the deeper regions of the earth under different conditions. Gas (mostly methane) may be produced by the deterioration of various types of organic matter (plants, etc.) or it may have been present from the origin of the earth. As regards the origin of any gas, the solar system gives convincing evidence of this, since the sun is known to be mostly composed of hydrogen gas. Gases inside the earth are present under high pressure and temperature. The chemical potential in the inner core ($\mu_{innercore}$) is higher than at the surface ($\mu_{surfaceearth}$) of the earth:

$$\mu_{innercore} > \mu_{surface\ earth} \tag{I.2}$$

The conventional reservoirs of oil and gas are thus an indication that the flow of oil/gas from the source rocks (i.e., shale) is based on the difference in the chemical potentials (Equation A.3):

$$\mu_{gas,shale} > \mu_{gas,\ surfaceofearth} \tag{I.3}$$

$$\mu_{oil,\ shale} > \mu_{oil,\ surface\ of\ earth} \tag{I.4}$$

where:

$\mu_{gas,shale}$ is the potential of gas in the shale reservoir
$\mu_{gas,surface\ of\ earth}$ is the potential of gas at the surface of the earth
$\mu_{oil,shale}$ is the potential of oil in the shale reservoir
$\mu_{oil,surface\ of\ earth}$ is the potential of oil at the surface of the earth

A drop in pressure is required to desorb the gas from the shale matrix (clay, organic matter, etc.).

Natural gas is a mixture of different gases, though the main component is methane (>90% as CH_4). It may contain other gases: butane, ethane, propane, and so on. Natural gas is known to burn cleanly and produces much lower harmful emissions (such as CO NOx, etc.) than coal or oil. Furthermore, it is known that the technology used in shale gas recovery is different from that used for conventional gas recovery (Ozkan, 1988; Burnham et al., 2006):

- Horizontal drilling
- Hydraulic fracturing
- Energy supply–demand urgency

Further, hydraulic fracturing was developed by using specific surface-active fracturing substances (SAFS). These substances have been known from experiments to induce fracture formation in both pure rock crystals and complex structures (such as cement).

Experiments have shown that the organic material in shale, kerogen, decomposes (by an endothermic process) at high temperatures. The temperature varies around 300°C. Since the oil is known to be strongly bound to the rock, it must be heated to make it flow for recovery. Oil shales can yield up to 100–200 L per ton of shale rock. The decomposition process is varied and complex. Hence, it will always move upward to the surface of the earth, where the pressure and temperature are lower. The same happened with oil, which resulted in its first recovery a century ago. Natural earthquakes and eruptions (and the subsequent creation of fractures) also lead to large movements of gases (and oil) from the inside of the earth. Further, kerogen (a fossil organic deposit) is reported to produce oil/gas on heating (Brandt, 2008; Nduagu and Gates, 2015). The shale-kerogen is heated (retorting process) in situ. Preliminary experiments show that this can be a useful process for oil/gas recovery. Mankind is essentially dependent on solar energy (especially the photosynthesis in plants), and the latter is expected to last for billions of years. Comparatively, the other energy sources are much more limited (at current estimates). Accordingly, the future aim for mankind is to find a means to survive on earth with energy sources that are at least as sustainable as the sun!

A1.1   COAL

Another related major source of energy is coal (Klimpel, 1995; Bondarenko et al., 2014; Speight, 2013). Some 200 million years ago, it is estimated that the atmospheric carbon dioxide levels were 1700–2000 ppm (i.e., five times today's 400 ppm). Over a geological history of about 200 (or more) million years, plants (among other things) that had been buried deep under the earth had been transformed under the influence of high pressure and temperature to organic materials (such as kerogen) (Calvin, 1969). These organic compounds were then converted to coal (solid), oil (liquid), or methane (gas). It may also be important to note that most of the carbon needed in this process had been taken from CO_2 in air. One can thus assert that when mankind burns coal, oil, methane, the CO_2 cycle in nature is being restored. At present, fossil fuels supply more than 90% of CO_2. The structure of coal is very much like a layer cake. Layers or seams are found to be separated from different inorganic substances. The thickness of seams varies from mine to mine by about 10 m. Coal mines can be many kilometers thick. The earth's CO_2 cycle is as follows:

CO_2 in air—Plants—Coal (carbonaceous products)

It is thus seen that nature converts CO_2 in air to plants, and later to carbon (coal). The carbon content in coal can vary from 25% to 85%.

Another observation of interest is that one finds rather large quantities of methane gas adsorbed in coal (Bumb and McKee, 1988). Methane is highly explosive and is the reason why explosions have occurred in some mines. It continues to desorb from coal long after it has left the mine, so the risk of explosion exists wherever coal is stored, especially in contained places. The most common method of dealing with desorbing methane is to vent it into the air. On the other hand, with 72 times the greenhouse effect of CO_2 over 20 years, this "liberated methane" is consequently the fourth largest anthropomorphic source of methane.

A1.2 COAL BURNING AND CO_2 PRODUCTION

$$Coal + Oxygen \rightarrow CO_2 \tag{I.5}$$

If one million tons of coal (carbon) is burned, then about 3.6 million tons of CO_2 is released. It is also found that virtually all the carbon in coal ends up in the atmosphere in the form of CO_2. In comparison, the burning of gasoline (as octane) in an oxygen atmosphere (or air) gives the equilibrium

$$2\left(C_8H_{18}\right) + 25\,O_2 \rightarrow 18\,H_2O + 16\,CO_2 \tag{I.6}$$

It is important to mention that CO_2 capture technology (CCT) is currently being developed. CCT essentially captures CO_2 (either in some suitable solvent or by converting CO_2 to other related products). Currently, some large CCT plants are actively employed worldwide.

Appendix II: Hydraulic Fracking Fluids (Surface Chemistry)

Geological analyses have shown that in shale reservoirs (source rock), methane (gas) is found in very large quantities. Gas is naturally trapped within highly fine-grained sedimentary rocks called *shale* (Calvin, 1969; Gudmundsson, 2000; Gupta and Hlidek, 2009; Shabro et al., 2012). Plant and animal debris decomposed (over millions of years) under high temperature and pressure into oil and gas. In some cases, oil and gas from shale migrated (through natural fractures) into the rock types toward the surface of the earth. These are called *conventional reservoirs*. Unconventional gas shales have shown that a different kind of technology is needed for such reservoirs, where gas is tightly bound (adsorbed). The shale rock matrix is made of very fine grains of minerals separated by very fine spaces, called *pores*. The gas in the shale matrix is found to be in two main distinct states:

- *Adsorbed* on the organic material
- *Free* in the numerous micropores

The pore space in shale rocks is known to vary from 2% to 10%. The porosity of rocks is a very important property for the oil/gas-recovery process. The chemical potentials of the two states are at equilibrium (according to classical thermodynamics) (Chattoraj and Birdi, 1984; Adamson and Gast, 1997; Somasundaran, 2015):

$$\mu_{\text{free,gas}} = \mu_{\text{adsorbed,gas}} \tag{II.1}$$

where:

$\mu_{\text{free,gas}}$ is the chemical potential of the adsorbed state
$\mu_{\text{adsorbed,gas}}$ is the chemical potential of the free gas state

The drilling of horizontal wells gives rise to enhanced gas release. The combination of hydraulic fracturing with this technology creates many kilometers of gas–rock surface area. As the gas pressure is released through the fractures to the production well, more gas desorbs, as required by the equilibrium constant (Equation II.1). The rate of this equilibrium depends on the shale structure and composition (i.e., total organic carbon [TOC], permeability, and ratio of inorganic: organic components).

The gas is known to be tightly bound to the shale rock, and it is known to be desorbed if there are fractures at low pressure. A few decades ago, a very effective method was applied to achieve this goal (so-called hydraulic fracturing) (Kale et al.,

2010; Gupta and Hlidek, 2009; Rogala et al., 2013). Hydraulic fracturing mainly involves using fluid (water) under high pressure, which creates multiple fractures or expands those that are already present in the matrix in the rock formation. At this stage, the mechanical force required for fracturing fluids (water) is used. However, other fracking fluid/gas mixtures (such as emulsions, foams, and CO_2/fluids) are also being investigated. The injection of fluids under high pressure induces fractures and fissures in the gas shale rocks. Shale reservoirs are found where the matrix is under stress from three dimensions. After this fracturing step, the fluid is removed for subsequent gas production (as determined by the desorption equilibrium). Fracture phenomena in solid matrix have been studied in the literature for decades (Lichtman et al., 1958; Liebowitz, 1971). Surface-active fracture substances (SAFS) are added to enhance the fracture process (Chapter 1). Among surface-active substances (SAS), a range of alcohols (methanol, isopropyl alcohol, 2-butoxy ethanol, and ethylene glycol) have been used to facilitate the fracking.

However, various other solid fracture phenomena (or cracking phenomena) have been reported in the literature. Some basic examples are

1. Glass cracking: mechanical process (after scratching with a diamond pen)
2. Metal cracking: surface molecular (aluminum crack formation after scratching with gallium)
3. Rock crystal under water: effect of charges (rock crystals after treatment with electrolytes [pH effect])
4. Shale rock (or similar): complex process (combination of 2 and 3) (SAFS additives: SAS or similar)

The adsorption of SAFS on shale rock will induce surface defects (on a molecular scale), which will lead to fracture formation. The fractures thus allow the gas (or oil) to flow to the wellbore. The rate of flow is dependent on the degree of fracturing as well as on the state of the gas (i.e., there is an equilibrium between adsorbed and free states). However, if one creates fractures in the shale reservoir, the gas in the free state will flow to the bore site.

The fracture technology has been known since 1947 in many different reservoir conditions (Ozkan, 1988; Ozkan et al., 2010; Zhang, 2002; Zhang and Gupta, 2002; Cahoy et al., 2013; Payman et al., 2014). Actually, over 600 fracture operations have already been reported between 1949 and 1954 in the United States alone (Clark, 1953; Cahoy et al., 2013; Smith and Montgomery, 2015; Zhang, 2002; Zhang and Gupta, 2002). The technology of hydraulic fracturing is one of the most important discoveries of this decade. Especially the combined horizontal drilling and fracture methods have had an enormous impact on oil/gas recovery. Based on this, there is increased interest in research and development. For example, the proppants used (such as resin coated) have been developed for deep wells. In fact, there are three major types of fracture techniques:

• Gel fracture
• Sand fracture
• Acid fracture

The hydraulic fracture fluids consist of various necessary additives. The most typical composition reported is

- Solvent (water) (from 90%–95%) is used to apply pressure in the horizontal boreholes. The high pressure gives rise to fractures.
- Fracture stabilizer: After mechanical force is applied to the reservoir (i.e., water at high pressure), a fracture is created. In order to stabilize this structure, very fine proppant is used (a granular material that prevents the created fractures from closing after the fracturing process). One finds different types of proppants: silica sand, resin-coated sand, bauxite, ceramics, or combinations of these. Since different shales exhibit different mechanical properties, the granular particles are selected accordingly (mostly dependent on characteristics of permeability or shale rock grain strength requirement). For example,
  - Low-pressure fractures use natural silica sand
  - High-pressure fractures may use higher-strength particles of bauxite or ceramics

The different fracturing developments of unconventional fluids have been concerned with various surface chemistry principles (Gregory, 2011; Gupta and Carman, 2011):

- Degree of fluid retention (related to the wetting properties of the shale).
- Fracturing fluid composition: water based, emulsions (mixtures of water and organic liquids [hydrocarbons]), foams (water + surfactants).
- Surfactant solutions with thixotropic viscosity (viscoelastic surfactant fluids): The solubility characteristics of surfactants change drastically with added electrolytes (Chapter 3). These characteristics give rise to some unusual viscosity effects (Birdi, 1997, 2016), foams (Chapter 7), and emulsions (Chapter 8).

Besides the fluid technology, the characteristic of proppants are also being investigated. The most commonly used proppant is silica sand. As mentioned in Chapter 1, the physical properties of colloidal particles are related to the size and shape of particles. It is reported that particles of uniform size and shape are believed to be more effective. This also suggests that more uniform-size proppant particles produce more uniform fractures.

Another investigation is concerned with the fate of the residual hydraulic water media in gas shale (Shabro, 2013; Howarth et al., 2011; Cahoy et al., 2013; Engelder et al., 2014). The application of horizontal drilling and high-volume hydraulic fracturing (HVHF) in this technology has been analyzed. Further, horizontal well technology has been applied since the 1930s in the United States. In a typical gas shale operation, some 10^4 m^3 of hydraulic water is injected into a horizontal well. The recovery from the well is less than 50%. The adsorption of additives under the ground is obviously of major interest as regards toxicity (Stringfellow and Domen, 2014).

Appendix II

A2.1 PERMEABILITY OF ROCKS (OIL/GAS RESERVOIRS)

The porous structure of both conventional and nonconventional rocks is of primary importance as regards oil/gas recovery. The flow of fluids through such porous solid material is measured by a method described in the literature (Bear, 1972; Walls, 1982; Ahmed, 2011; Birdi, 1999, 2016). The quantity permeability constant of a porous material, k_{Darcy}, is defined as the relation between the flow rate of the fluid, V_{flow}, the permeability constant, k_{Darcy}, the viscosity of the fluid, u_{fluid}, the pressure difference, dP, and the thickness of the solid material, dx:

$$V_{flow} = \left(k_{Darcy} dP \right) / \left(u_{fluid} dx \right) \tag{II.2}$$

For example, a porous solid with a permeability constant of 1 D has the following data:

- $V_{flow} = 1 \text{ cm}^3/\text{s}$
- $\mu_{fluid} = 1 \text{ cP (1 mPa s)}$
- $dP = 1 \text{ atm/cm (across a 1 cm}^2 \text{ area)}$

Typical values of permeability constants, k_{Darcy}, for different porous rocks are reported as

Gravel: 100,000 D
Sand: 1 D
Granite: <0.01 μD

Appendix III: Effect of Temperature and Pressure on Surface Tension of Liquids (Corresponding States Theory)

In both industry and research, one manipulates a large amount of data, which could be systemized. According to the chemistry and physics of liquid surfaces, it is thus important to be able to describe the interfacial forces as a function of temperature and pressure. This is of particular interest in all oil and gas reservoirs, which are generally found at high temperatures and pressures (Birdi, 2003, 2010b, 2016; Somasundaran, 2015). In the case of all liquids, the magnitude of a γ decreases almost linearly with temperature within a narrow range (Partington, 1951; Birdi, 2003; Defay et al., 1966):

$$\gamma t = k_0 \left(1 - k_1 t \right) \tag{III.1}$$

where k_0 is a constant. It was found that coefficient k_1 is approximately equal to the rate of decrease of density (ρ) with the rise in temperature:

$$\rho t = \rho_0 \left(1 - k_1 t \right) \tag{III.2}$$

Values of constant k_1 were found to be different for different liquids. Furthermore, the value of γ was related to critical temperature (T_C).

Equation III.3 relates the surface tension of a liquid to the density of liquid, ρ_1, and vapor, ρ_v (Partington, 1951; Birdi, 1989):

$$\gamma / \left(\rho_1 - \rho_v \right)^4 = C_{constant} \tag{III.3}$$

where the value of constant $C_{constant}$ is nonvariable only for organic liquids, while it is not constant for liquid metals.

At the critical temperature, T_C, and critical pressure, P_c, a liquid and its vapor are identical, and the surface tension, γ, and total surface energy, like the energy of vaporization, must be zero (Partington, 1951; Birdi, 1997). At temperatures below the boiling point, which is 2/3 T_C, the total surface energy and the energy of evaporation are nearly constant. The variation in surface tension, γ, with temperature is given in Table A3.1 for different liquids.

This data clearly shows that the variation of γ with temperature is a very characteristic physical property (the slope is the surface entropy: see Equation III.8). This

TABLE A3.1

Variation of γ with Temperature of Different Alkanes

n-Alkane	Temperature (°C)	Measured γ	Calculated γ
C_5	0	18.23	18.25
	50	12.91	12.8
C_6	0	20.45	20.40
	60	14.31	14.3
C_7	30	19.16	19.17
	80	14.31	14.26
C_9	0	24.76	24.70
	50	19.97	20.05
	100	15.41	15.4
C_{14}	10	27.47	27.4
	100	19.66	19.60
C_{16}	50	24.90	24.90
C_{18}	30	27.50	27.50
	100	21.58	21.60

observation becomes even more important when it is considered that the sensitivity of γ measurements can be as high as approximately ±0.001 dyne/cm (=mN/m). The change in γ with temperature in the case of mixtures would thus be dependent on the composition. The addition of gas to a liquid always decreases the value of γ. For example, the variation of γ of the system CH_4 + hexane is given as

$$\gamma(CH_4 + \text{hexane}) = 0.64 + 17.85 \, x_{\text{hexane}} \qquad \text{(III.4)}$$

It is seen that by measuring γ for such a system, one can actually estimate the concentration of CH_4. This is of great interest for oil reservoir engineering operations, in which one invariably finds CH_4 in crude oil.

- Methane: CH_4
- Boiling point: −161.5°C
- Density: 0.656 kg/m³
- Melting point: −182°C

It is well known that the corresponding states theory can provide much useful information about the thermodynamics and transport properties of fluids. For example, the most useful two-parameter empirical expression, which relates the surface tension, γ, to the critical temperature, T_C, is given as

$$\gamma = k_0 \left(1 - T / T_C\right)^{k_1} \qquad \text{(III.5)}$$

where k_0 and k_1 are constants. Van der Waals derived this equation and showed that $k_1 = 3/2$, although experiments indicated that $k_1 = 1.23$. Guggenheim (Partington, 1951) has suggested that $k_1 = 11/9$. However, for many liquids, the value of k_1 lies between 6/5 and 5/4.

It was found that k_0 was proportional to $T_C^{1/3} P_c^{2/3}$. Equation III.5, when fitted to the surface tension, γ, data of liquid CH_4, has been found to give the relationship

$$\gamma CH_4 = 40.52 \left(1 - T/190.55\right)^{1.287} \tag{III.6}$$

where $T_C = 190.55$ K. This equation has been found to fit the γ data for liquid methane from 91 to 190°K, with an accuracy of ± 0.5 mN/m.

A range of γ versus T data on n-alkanes, from n-pentane to n-hexadecane, has been analyzed (Birdi, 1997a, 1997b). The constants k_0 (between 52 and 58) and k_1 (between 1.2 and 1.5) were found to be dependent on the number of carbon atoms, Cn. The estimated values from Equation III.8 for different n-alkanes were found to agree with the measured data within a few percent: γ for n-$C_{18}H_{38}$, at 100°C, was 21.6 mN/m from both measured and calculated values. This agreement shows that the surface tension data on n-alkanes fits the corresponding state equation very satisfactorily. It is worth mentioning that the equation for the data on γ versus T, for polar (and associating) molecules such as water and alcohols, when analyzed by Equation III.5, gives magnitudes of k_0 and k_1 that are significantly different from those found for nonpolar molecules such as alkanes. This difference indicates that the change with temperature of surface forces is different (Birdi, 1997).

The variation of γ for water with temperature (t/°C) is given as (Cini et al., 1972; Birdi, 1997)

$$\gamma H_2O = 75.668 - 0.1396\, t - 0.2885 \times 10^{-3} t^2 \tag{III.7}$$

The surface entropy (S_s) corresponding to Equation III.5 is

$$S_S = -d\gamma/dT \tag{III.8}$$

and the corresponding surface enthalpy, h_s

$$h_s = g_s - Ts_s$$

$$= -T\,(d\gamma/dT) \tag{III.9}$$

The reason heat is absorbed on expansion of a surface is that the molecules must be transferred from the interior against the inward attractive force, to form the new surface. In this process, the motion of the molecules is retarded by this inward attraction, so that the temperature of the surface layers is lower than that of the interior, unless heat is supplied from outside. Further, extrapolation of γ to zero surface tension in the data gave values of T_C that were 10%–25% lower than the measured values (Birdi, 1997, 2003). This deviation was recently analyzed in great detail.

The surface tension, γ, of any liquid would be related to the pressure, P, as follows:

$$(d\gamma/dP)_{A,T} = (dV/dA)_{P,T} \tag{III.10}$$

Since the quantity on the right-hand side would be positive, then the effect of pressure should be to give an increase in γ. Preliminary analyses indicate that the term (dγ/dP) is positive and dependent on the alkane chain length (Jennings, 1967; Birdi, 1997). The data for benzene (C_6H_6) at 20°C is

- $(d\gamma/dT)_{20\,C-30\,C} = -0.04$ mN/m/°C
- $(d\gamma/dP)_{25\,C,\,1-50\,atm} = 0.07$ mN/m/atm

This data is important in many systems where high pressure is present, for example:

- Oil reservoirs (100–200 atm pressure)
- Car tires (exert high pressure on the roads)
- Teeth (exert considerable pressure)
- Shoes (exposed to high pressures)
- Building structures

The surface tension data of mixtures of methane (CH_4) and nonane (C_9H_20) has been analyzed as a function of pressure (10–80 atm) (Deam and Mattox, 1970; Reid et al., 1987). It was found that at a given temperature, the magnitude of surface tension decreased with increasing pressure (this is indicative of higher solubility of methane). Similar studies have also been reported for methane–pentane and methane–decane mixtures (Stegemeier, 1959).

The following relationship relates surface tension, γ, of liquids to the density (Reid et al., 1987; Birdi, 1997):

$$\gamma(M_w/\rho)^{2/3} = k \left(T_c - T - 6\right) \tag{III.11}$$

where:

M_w is the molecular weight
ρ is the density (M_w/ρ = molar volume)

The quantity ($\gamma (M_w/\rho)^{2/3}$) is called the *molecular surface energy*. It is important to notice the correction term 6 on the right-hand side. This is the same as is found for n-alkanes and n-alkenes in the estimation of T_C from γ versus temperature data (Birdi, 1997).

Calculated γ (Birdi, 1997) and measured values of different n-alkanes at various temperatures are shown in Table A3.1.

Appendix IV: Solubility of Organic Molecules in Water: A Surface Tension—Cavity Model System (Structure of Water and Gas Hydrates)

In all solutions, the solubility property is of primary interest. Water is the most important liquid solvent in everyday life. More than 70% of earth is covered with oceans (plus lakes and rivers). Hence, the solubility characteristics of all kinds of molecules in water are of major importance in technology and different phenomena. The mechanism of solubility in water of an inorganic salt, such as NaCl, is found to be different from that of an alkane (such as hexane, pentane, butane, propane, ethane, or methane) molecule. In the aqueous phase, NaCl dissociates into Na^+ and Cl^- ions and interacts with water through hydrogen bonds. A hexane molecule merely dissolves (although with very low solubility) in water by entering inside the water structure. Since the water structure is stabilized mainly by hydrogen bonds, the hexane molecule will give rise to some rearrangements of these bonds, but without breaking the hydrogen bonds. This conclusion is based on the fact that no heat is evolved in the solution process (Tanford, 1980; Birdi, 2016). It has been suggested that organic molecules (such as alkanes) dissolve in water by creating a cavity. This model was found to be able to predict the solubilities of both simple and more complicated organic molecules. In the simplest case, the solubility of heptane is lower than that of hexane, due to the addition of one $-CH_2-$ group. In the case of alkane molecules, a linear relation is found between the solubility and the number of $-CH_2-$ groups (Tanford, 1980; Birdi, 1997).

This model is thus based on the following assumptions when an alkane molecule (deoicted as CCCCCCCCC) is placed in water (depicted as w):

- wwwwwwwwwwwwwwwwwww
- wwwCCCCCCCCCwwwwwwww
- wwwwwwwwwwwwwwwwwww

The energy needed to create the surface area of the cavity will be proportional to the degree of solubility of the alkane. Thus, the solubility of any alkane molecule will be given as

Free energy of solubility

= proportional to the(cavity surface area)(surface tension of the cavity)

By analyzing the solubility data of a whole range of alkane molecules, the following relation was found to fit the experimental data:

Free energy of solubility = ΔG°_{sol}

$$= RT\,Log(solubility) \tag{IV.1}$$

$$= (\gamma_{cavity})(S_{areaalkane})$$

$$= 25.5(S_{areaalkane}) \tag{IV.2}$$

where $S_{areaalkane}$ is the area of the alkane cavity. For the solubility of linear alkanes in water, the total surface area (TSA) gives the solubility

$$\ln(sol) = -0.043(TSA) + 11.78 \tag{IV.3}$$

where solubility is in molar units and TSA in Ångstroms squared. For example,

Alkane	Solubility	TSA	Predicted (sol)	Ratio
n-Butane	0.00234	255	0.00143	1
n-Pentane	0.00054	287	0.0004	1/4.3
n-Hexane	0.0001	310	0.0001	1/5.4
n-Butanol	1.0	272	0.82	1
n-Pentanol	0.26	304	0.21	1/4
n-Hexanol	0.06	336	0.05	1/4

This data shows that the addition of each $-CH_2-$ group in alkanes or alcohols reduces the solubility in water by a factor of ca. 4. The constant 0.043 in Equation IV.3 (Tanford, 1980) is estimated to be equal to $\gamma_{cavity}/RT = 25.5/600$. The surface areas of each group in n-nonanol were estimated by different methods (from molecular models, geometrical areas, and computational methods). These estimated surface areas (in $Å^2$) are as follows:

CH_3	CH_2	CH_2	CH_2	CH_2	CH_2	CH_2	CH_2	CH_2	OH
85	43	32	32	32	32	32	40	45	59

It is seen that if one needs to estimate the value of TSA of n-decanol, it will be (TSA of nonanol + TSA of CH_2), that is, 431 + 32 = 463 $Å^2$.

The data for solubility of a homologous series of n-alcohols is given here.

Alcohol	Solubility(mol/L)	Log(S)
C_4OH	0.97	–0.013
C_5OH	0.25	–0.60
C_6OH	0.06	–1.22
C_7OH	0.015	–1.83
C_8OH	0.004	–2.42
C_9OH	0.001	–3.01
$C_{10}OH$	0.00023	–3.63

Accordingly, this algorithm allows one to estimate the solubility in water of any organic substance with known structure. The estimated solubility of cholesterol ($C_{27}H_{46}O$; a rather large, complex molecule with very low solubility: 1.8×10^{-6} g/mL = 1.8 mg/mL) was almost in accord with the experimental data (Tanford, 1980; Birdi, 2003). It is seen that log(S) is a linear function of the number of carbon atoms in the alcohol. Each $-CH_2-$ group reduces log(S) by 0.06 unit.

A4.1   STRUCTURE OF WATER (ICE) AND GAS HYDRATES

The model of water structure described in regard to the solubility of hydrocarbons needs some more detailed analysis. The structure of solids is studied by using methods that provide a detailed molecular structure (e.g., x-ray diffraction). This is obviously not possible in the case of liquids. Water exhibits some properties that are not normal for liquids. For example, all normal liquids on cooling show an increase in density till their freezing point is reached. However, water exhibits a maximum density at 4°C (at 1 atm pressure). Heavy water, D_2O, exhibits a maximum density at 11.2°C (Venkateswarlu, 1996; Brezonik and Arnold, 2011). The maximum in density is interpreted as arising from some kind of structure change at this temperature. The water becomes less dense between 4°C and freezing point (0°C). This is why ice is ca. 10% less dense than water in the liquid state (ice floats on water).

The cagelike water structure in ice suggests some geometrical packing and its properties (Figure A4.1a). This structure will also be present with fluctuations in the liquid water. The distance between water molecules is found to be sufficient to accommodate a methane molecule (Figure A4.1b,c). The different hydrate complexes are (Tanford, 1980; Sloan and Koh, 2007; Aman, 2016):

- Methane hydrate: $CH_4-\{5 (3/4)\}$ H_2O
- Chlorine hydrate: $Cl_2-\{8\}$ H_2O

These data show that in ice there is a complex structure whereby 5(3/4) water molecules are arranged such that one molecule of CH_4 can be placed within. Similarly, the Cl_2 molecule can form a complex whereby in ice, eight water molecules are surrounding the hydrate (Figure A4.1c). These hydrate structures are useful in explaining the solubility of nonpolar organic substances in water. Especially, the surface area solubility model has been very useful in this context (Tanford, 1980; Birdi, 2016).

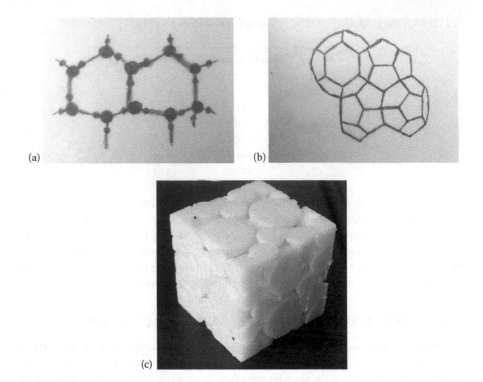

(a)   (b)

(c)

FIGURE A4.1   (a) The cage-like structure of ice. (b) A schematic gas-hydrate structure (only water molecules are indicated). (c) One cell model (3D printed by Ultimaker) of a gas-hydrate: water (as small molecules) and chlorine (as large molecules). (From thingiverse.com/thing:1371876.)

Furthermore, a very large accumulation of methane as hydrate is known to exist on earth, mainly offshore near the continental sediment region, as well as in polar regions (the permafrost region) (Kvenvolden, 1995; Hesselbo et al., 2000). It is suggested that large amounts of methane exist in the inner core of the earth. This gas has migrated, and some of it is trapped in the ice as hydrate.

Appendix V: Gas Adsorption–Desorption on Solid Surfaces

A clean solid surface is an active site for the adsorption of gas or vapor molecules. Adsorption on solid surfaces is of fundamental importance in a range of industrial and everyday phenomena. The amount of adsorption of a gas on a solid is found to be related to

- Temperature
- Gas vapor pressure
- Solid surface area available for adsorption (m²/g)

This also indicates that to understand the process, the magnitude of the solid surface is needed. There exist different forms of surface forces between the gas and the solid. The attraction forces between a gas and a solid are mainly the van der Waals type (Jaycock and Parfitt, 1981; Adamson and Gast, 1997; Birdi, 2010b, 2016; Somasundaran, 2015). This suggests that there is a relation between the critical temperature, T_C, of a gas and the degree of adsorption on a solid. It is found that greater T_C leads to greater adsorption:

Gas	H_2	N_2	CO	CH_4	CO_2	NH_3
T_C (K)	33	126	134	190	304	406

The process of gas adsorption by solid surfaces is relevant to both industrial (gas reservoirs, purification and drying of gases, solvent recovery, fractional separation and capture (CO_2), gaseous reactions catalyzed by solid surfaces [formation of ammonia (NH_3) from N_2 and H_2], etc.) and other processes (pharmaceutical, etc.) (Jaycock and Parfitt, 1981).

At equilibrium, gas molecules will adsorb to and desorb from the solid surface at the same rate; the rates of adsorption (R_{ads}) and desorption (R_{des}) will be equal. The surface can be described as consisting of (Chattoraj and Birdi, 1984; Adamson and Gast, 1997; Somasundaran, 2015; Birdi, 2016)

- Total surface area $= A_t = A_o + A_m$
- Area of clean surface $= A_o$
- Area covered with gas $= A_m$
- Enthalpy of adsorption $= E_{ads}$

One can write

$$R_{ads} = k_a p A_o \tag{V.1}$$

$$R_{des} = k_b A_m \exp\left(-E_{ads} / RT\right) \tag{V.2}$$

where k_1 and k_2 are constants.
   At equilibrium,

$$R_{ads} = R_{des} \tag{V.3}$$

and the magnitude of A_o is constant.
   Further, one has

Amount of gas adsorbed $= N_s$

Monolayer capacity of the solid surface $= N_{sm}$

By combining these relations and

$$N_s / N_{sm} = A_m = A_t \tag{V.4}$$

one obtains the well-known Langmuir adsorption equation:

$$N_s = N_{sm} / \left(ap\right) / \left(1 + \left(ap\right)\right) \tag{V.5}$$

Additionally, the heat of adsorption has been investigated. For example, the amount of Kr gas adsorbed on AgI increases when the temperature is decreased from 79 K (0.13 ml/g) to 77 K (0.16 mL/g). This data allows one to estimate the isosteric heat of adsorption (Jaycock and Parfitt, 1981):

$$\left(d\left(Ln\ P\right) / dT\right) = q_{ads} / RT^2 \tag{V.6}$$

The measured magnitudes of q_{ads} were in the range of 10–20 kJ/mol.
   Since the adsorption process is found to be a spontaneous phenomenon, this means that the free energy $\Delta G < 0$ (i.e., negative enthalpy)

$$\Delta G = \Delta H - T\Delta S \tag{V.7}$$

Since the adsorption process means a loss in entropy, $\Delta S < 0$, the sign of ΔH must be negative (i.e., <0). This means that the adsorption process is exothermic. Thus, the degree of adsorption increases with decreasing temperature (and decreases with increasing temperature).
   Another example of adsorption of N_2 (at 77.3 K) on alumina (Al) powder shows

Pressure (P/mm Hg)	V (=g N₂/g alumina)
31.7	0.000831
56.6	0.000890
112.4	0.001015
169.3	0.001118

From this data, the area per molecule of adsorbed N_2 was found to be 16.2 A. This value is in reasonable agreement with other methods.

Appendix VI: Common Physical Fundamental Constants

- Angstrom, $\mathring{A} = 10^{-8}$ cm $= 10^{-10}$ m
- Micrometer $= \mu$m $= 10^{-6}$ m
- nm $= 10^{-9}$ m $= 10\,\mathring{A}$
- Boltzmann constant (R/N_A), $k_B = 1.381 \times 10^{-23}$ J/K
- Electronic charge (F/N_A), $e = 1.602 \times 10^{-19}$ C
- Avogadro's constant $= N_A = 6.023 \times 10^{23}$ molecules/mole
- Faraday constant $(F) = 96{,}496$ coulomb equiv.$^{-1}$
- Gas constant $(R) = 8.314$ J/K mole $= 1.986$ cal/K mole/
- Ideal gas law $(P = $ pressure; $V = $ volume$)$ $T = 4.12 \times 10^{-22}$ J at 298 K (25°C)
- Molar gas volume (at normal conditions) $= V_m = 22.4 \times 10^{-3}$ m^3/mol (22.4 L/mole/)
- 1 cal $= 4.184$ J
- 1 erg $= 10^{-7}$ J
- 1 atm $= 1.013 \times 10^5$ N/m^2 (Pa)
- kT/e $= 25.7$ mV (at 25°C)
- Permittivity of free space, $e_o = 8.854 \times 10^{-12}$ C$_2$/J/m
- eV $= 1.6021 \times 10^{-19}$ J
- Kilowatt hour $=$ kWh $= 3.6 \times 10^6$ J
- Viscosity of water $= 0.001$ N s/m$^2 = 1$ centipoise, at 20°C
- Dielectric constant of water $= 80.2$ at 20°C
- Gravity constant $= g = 9.81$ m/s^2
- Speed of light $= 2.997 \times 10^8$ m/s $= 300$ km/s

Index

A

Abrading process, 128
Adhesive, 127–129
  and adhesion, 97–98
  failures, 129
Adsorption–desorption
  energy, 10, 11
  process, 119
Adsorption/desorption, on solid surfaces, 98–113
  area determination, 110–113
  gas, on measurement methods, 105–107
    gravimetric, 106
    Langmuir, 106–107
    volumetric change, 105–106
  of solutes from solution on, 109–110
  various gas, analyses, 107–109
AFM, *see* Atomic force microscope (AFM)
Amphiphile, 57, 59
  aqueous solutions of surface-active
    substances, 60–62
  orientation of, molecules at oil–water
    interfaces, 178
Anionic surfactants, 65
Antifoaming agents, 157, 159
Antonow rule, 52
AOT, *see* Sodium bis-(2-ethylhexyl)
    sulfosuccinate
Atomic force microscope (AFM), 146, 164–166
Attractive, and repulsive forces, 135–137,
    164–166

B

BET, *see* Brunauer–Emmett–Teller (BET) model
Binnig, G., 163
Biodegradation, 124
Black film state, 155
Black shale, 193
Boltzmann distribution, 138
Brownian motion, 131
Brunauer–Emmett–Teller (BET) model, 104,
    108–109
Bubbles, and foams, 151–166
  application of, in technology, 152
  applications of scanning probe microscopes
    to surface and colloid chemistry,
    161–166
    measurement of attractive and repulsive
      forces by AFM, 164–166

foams, 153–159
  antifoaming agents, 159
  formation and surface viscosity, 158–159
  stability, 155–158
  wastewater purification, 159–161
    froth flotation and bubble foam methods,
      160–161
Bypass phenomenon, 120

C

Calorimetric measurements, 114
Capillary forces, in fluid flow, 21–53
  bubble/foam formation, 36–38
  interfacial tension of liquid$_1$ (oil)–liquid$_2$
    (water), 51–53
    measurement of, 52–53
  Laplace equation, 27–33
  rise/fall, 33–36
  surface forces, 23–26
    energy, 24–26
  surface tension
    data, 43–46
    effect of temperature and pressure on,
      46–51
    measurement of, 38–43
Carbon black, 103
Cetyl trimethyl ammonium bromide (CTAB), 68
Chemistry, of detergency, 124–126
Cloud point (CP), 64
Coagulation
  flocculation and, of colloidal suspension,
    146–147
  kinetics of, colloids, 145–146
Coal, 199, 200
Cohesive failure, 129
Colloidal systems, hydraulic fracking technology,
    15, 17–18, 131–149
  Derjaguin–Landau–Verwey–Overbeek
    (DLVO) theory, 134–142
  silica suspension in, 134–142
  stability of lyophobic suspensions, 142–147
  wastewater treatment and control, 147–149
Contact angle (θ)
  of liquids on solid surfaces, 94–95
  measurements of, at liquid–solid interfaces,
    95–97
Conventional reservoirs, 2–3, 195, 201
Corresponding states theory, 205–208

219

CO_2 production, coal burning and, 200
Coulter counter, 175
CP, *see* Cloud point (CP)
Crack propagation, 9
Critical concentration, 143
Critical micelle concentration (CMC), 65, 68, 69,
    70, 71, 83, 85
Crystalline/lamellar phase, 173
CTAB, *see* Cetyl trimethyl ammonium
    bromide (CTAB)

D

Darcy's law, 101
Debye-Huckel (D-H) theory, 139
Derjaguin–Landau–Verwey–Overbeek (DLVO)
    theory, 134–142, 146
    charged colloids, 137–141
    electrokinetic processes of charged particles
        in liquids, 141–142
Detachment method, 40
Detergents, 56, 124–126
Diffusion bonding, 128
Diffusion process, of gas, 99, 101
Direct interactive force, 164–166
DLVO, *see* Derjaguin–Landau–Verwey–
    Overbeek (DLVO) theory
Drop size analyses, emulsion, 175

E

EDL, *see* Electrical double-layer (EDL)
Electrical charge distribution, 137–141
Electrical double-layer (EDL), 153,
    176, 177
Electrokinetic processes, of charged particles,
    141–142
Electro-osmosis, 141
Electrophoresis, 141, 142
Electrostatic forces, 128
Emulsions, 18–19, 167–178
    and hydraulic fracking, 168
    stability and analyses, 175–178
        creaming/flocculation of drops,
            177–178
        electrical, stability, 176–177
    structure of, 168–175
        HLB values of emulsifiers, 170–173
        methods of, formation, 173–175
        oil–water, 169–170
Enhanced oil recovery (EOR), 181
EOPO, *see* Ethylene oxide–propylene oxide
    (EOPO)
EOR, *see* Enhanced oil recovery (EOR)
Epoxy, 128
Ethylene oxide–propylene oxide (EOPO), 125
Evaporation rates, of liquid drops, 126–127

F

FFM, 164
Flocculation, and coagulation of colloidal
    suspension, 146–147
Flotation, 115–117, 159
Flow-back, 92
Flow of gas, 10
    in porous media (shale), 10, 13, 100
Foams, 153–159
    antifoaming agents, 159
    application of, and bubbles in
        technology, 152
    formation and surface viscosity, 158–159
    stability, 155–158
Fracking, 7, 8, 16, 46, 55
Fracture
    formation, 9, 14
        mechanism of, 9
        minimum stress, 113
    hydraulic, fluid injection and wettability of
        shales, 92–94
    stabilizer, 203
    surface phenomena in solid-adsorption and,
        process, 113
    systems, 14–15
Friction, 115
Froth flotation, and bubble foam methods,
    160–161

G

Gas adsorption
    analyses, 107–109
    –desorption, on solid surfaces, 213–215
    on solid measurement methods, 105–107
        gravimetric, 106
        Langmuir, 106–107
        volumetric change, 105–106
    on solids
        adsorption–desorption (hysteresis), 108
        enthalpy, 108
        molecular spacing, 104
Gas desorption, 8
Gas-recovery mechanism, 197
Geochemistry
    of shale gas reservoirs, 191–200
    surface chemistry and, of hydraulic
        fracturing, 1–19
        colloids, 17–18
        emulsions, 18–19
        fluids, 18–19
        formation of fractures in shale reservoirs
            and surface forces, 7–17
Gibbs adsorption
    equation, 72–83
    theory, 16–17

Gravimetric gas adsorption methods, 106
Gushing process, 158, 159

H

Hand soap, 173
Heat of adsorption (of gas), 104
High-pressure injection, of water, 5, 7
High-volume hydraulic fracturing (HVHF), 203
HLB, *see* Hydrophilic–hydrophobic balance (HLB)
Horizontal drilling, 203
HVHF, *see* High-volume hydraulic fracturing (HVHF)
Hydraulic fracking, 2, 115–117
    emulsions and, 168
    fluids, 18–19, 201–204
    technology, 131–149
        Derjaguin–Landau–Verwey–Overbeek (DLVO) theory, 134–142
        silica suspension in, 134–142
        stability of lyophobic suspensions, 142–147
        wastewater treatment and control, 147–149
    wetting, 89
Hydraulic fracturing
    fluid injection and wettability of shales, 92–94
    surface chemistry and geochemistry of, 1–19
        colloids, 17–18
        emulsions, 18–19
        fluids, 18–19
        formation of fractures in shale reservoirs and surface forces, 7–17
Hydrophilic–hydrophobic balance (HLB), 57, 58, 92, 170–173

I

IEP, *see* Isoelectric point (IEP)
IFT, *see* Surface tension
Interfacial chemistry, 5
Interfacial tension (IFT), *see* Surface tension
Ionic surfactants, 62–64
Isoelectric point (IEP), 165

K

Kerogen, 196
Kinetics, of coagulation, 145–146
Knudsen diffusion, 100
Knudsen diffusion domain, 4
Knudsen number, 99
KP, *see* Krafft point (KP)
Krafft point (KP), 63–64
Kugelschaum, 154

L

Langmuir adsorption isotherm, 110
Langmuir adsorption (of gas), 103, 107
Langmuir equation, 107, 108, 111
Langmuir gas adsorption, 106–107
Laplace equation, for liquids, 27–33
Laser light-scattering instruments, 175
Liquid crystals (LC), 168
Liquid drop weight, and shape method, 38–41, 52
    maximum weight method, 40
    pendant drop method, 40–41
Liquid surface
    curved, 22
    flat, 22
Liquid surface formation, and evaporation, 48–51
London potential, 135
Lyophobic suspensions, stability of, 142–147
    flocculation and coagulation of colloidal suspension, 146–147
    kinetics of coagulation of colloids, 145–146
Lyotropic LCs, 168

M

Macroscopic technology, 3
Matter, structure of, 5–6
Maximum weight method, 40
Mechanical interlocking, 128
Methane, 199
Micelle
    formation of surfactants, 65–72
    solubilization in, 83–86
Microemulsions, 168, 178–182
    detergent, 180–181
    technology for oil reservoirs, 181–182
Microporous rocks, 99
Microscopic analyses, 3
Monodisperse, 134
Multilayer gas adsorption, 108

N

Nano-reactors, 83
Nonconventional sources, 2, 3
Nonionic surfactants, 64–65

O

Oil, and water
    emulsions, 18, 19, 169–170
    orientation of amphiphile molecules at, interfaces, 178
Oil/gas reservoirs, 204
Oil reservoirs, microemulsion technology for, 181–182

Oil spills, on oceans
  and clean-up process, 122
  different states of, 122–124
Organic molecules, solubility of, 209–212
Organic water-insoluble molecules, solubilization
  of, 83–86

P

Pendant drop method, 40–41
Permeability, of rocks, 204
Phospholipids, 57
Plate method, Wilhelmy, 41–43
Polishing, 115
Polydisperse, 134
Polystyrene (PS), 95, 96
Porosity, 197, 201
Porphyrins, 194
Proppant, 203
PS, *see* Polystyrene (PS)

Q

Quantitative structure activity related (QSAR)
  analyses, 23

R

Rocks, permeability of, 204
Rohrer, H., 163

S

SAFS, *see* Surface active fracture substances
  (SAFS)
Saponification, 58
SAS, *see* Surface active substances
Scanning probe microscopes (SPM), applications
  of, 161–166
Scanning tunneling microscope (STM),
  163, 165
SDS, *see* Sodium dodecyl sulfate (SDS)
Sedimentation potential, 141
Self-assembly, 91
Shale gas reservoirs
  adsorption and desorption, 98–105
  formation of fractures in, 7–17
  geochemistry of, 191–200
Shales, 201
  hydraulic fracture fluid injection and
    wettability of, 92–94
  rock and other solid surfaces, 164–166
Soaps, 56
Sodium bis-(2-ethylhexyl) sulfosuccinate, 180
Sodium dodecyl sulfate (SDS), 68, 77, 80, 83
Solid surface
  gas adsorption–desorption on, 213–215

wetting, adsorption, and related processes,
    119–129
  adhesion phenomena, 127–129
  chemistry of detergency, 124–126
  evaporation rates of liquid drops, 126–127
  oil and gas recovery and surface forces,
    120–124
Solid surfaces, surface chemistry of, 87–117
  adsorption/desorption on, 98–113
    area determination, 110–113
    gas, on measurement methods, 105–107
    of solutes from solution on, 109–110
    various gas, analyses, 107–109
  contact angle (θ) of liquids on, 94–95
  degree of surface roughness, 115
  friction, 115
  heats of adsorption on, 113–114
  measurements of contact angles at liquid–
    solid interfaces, 95–97
  phenomena in solid-adsorption and fracture
    process, 113
  phenomena of flotation, 115–117
  surface tension of, 94
  theory of adhesives and adhesion, 97–98
  wetting properties of, 89–94
    hydraulic fracture fluid injection and
      wettability of shales, 92–94
Solubility, of organic molecules, 209–212
Solvent, 203
SPM, *see* Scanning probe microscopes (SPM)
Square shaped capillary, 121
Stern/Helmholtz layer, 145
STM, *see* Scanning tunneling microscope (STM)
Straight-chain alkanes, 48
Streaming potential, 141, 145
Super-hydrophobic surfaces, 96
Surface, and colloid chemistry, 161–166
Surface active, and fracture-forming substances
    (SAS and SAFS), 55–86, 202
  Gibbs adsorption equation in solutions,
    72–83
    kinetic aspects of surface tension of
      detergent aqueous solutions, 81–83
  micelle
    formation of surfactants, 65–72
    solubilization in, 83–86
  surface tension of aqueous solutions, 58–65
    amphiphiles, 60–62
    solubility characteristics of surfactants in
      water, 62–65
Surface active fracture substances (SAFS), 5, 8,
    9, 14, 113, 198, 202
Surface chemistry
  and geochemistry of hydraulic fracturing,
    1–19
    colloids, 17–18
    emulsions, 18–19

fluids, 18–19
formation of fractures in shale reservoirs and surface forces, 7–17
hydraulic fracking fluids, 201–204
of solid surfaces, 87–117
adsorption/desorption on, 98–113
contact angle (θ) of liquids on, 94–95
degree of surface roughness, 115
flotation, phenomena of, 115–117
friction, 115
heats of adsorption on, 113 114
measurements of contact angles at liquid–solid interfaces, 95–97
phenomena in solid-adsorption and fracture process, 113
surface tension of, 94
theory of adhesives and adhesion, 97–98
wetting properties of, 89–94
Surface diffusion, 4
Surface entropy, of liquids, 47, 48
Surface forces, 4, 7–17, 22, 23–26
Surface roughness, degree, of, 115
Surface tension, 11, 12, 13, 16, 25, 51–53, 56, 72, 73, 81, 169
of aqueous solutions, 58–65
kinetic aspects of, detergent, 81–83
solubility characteristics of surfactants in water, 62–65
surface-active substances, 60 62
effect of temperature and pressure on, of liquids, 205–208
of liquids, 25, 27
data, 43–46
effect of temperature and pressure on, 46–51
measurement of, 38–43
of solids, 94
Surface tension—cavity model system, 209–212

Surfactants, 56, 90–91
micelle formation of, 65–72
solubility characteristics of, in water, 62–65

T

Tertiary oil recovery, 120
Thin liquid films (TLF), 151, 153, 155, 157
Total organic carbon (TOC), 193
Total organic content (TOC), 92
Tribology, 115

V

Van der Waals forces, 133, 135, 136
Volumetric change methods, gas, 105–106

W

Wastewater
purification, 159–161
froth flotation and bubble foam methods, 160–161
treatment and control, 132, 133, 147–149
Water, and gas hydrates, 211–212
Water-by-pass
oil recovery, 121
Water-in-oil emulsions, 170
Wetting, 21, 89–94, 119
Wicking process, 33
Wilhelmy plate method, 41–43, 52

Y

Young's equation, 91, 92, 96
Young's modulus, 113

Z

Zeta potential, 148

Milton Keynes UK
Ingram Content Group UK Ltd.
UKHW040102071024
449327UK00019B/759